MONOGRAPHS
ON STATISTICAL SUBJECTS

═══════

General Editor: M. S. Bartlett

QUEUES

QUEUES

D. R. COX
Imperial College,
University of London

AND

WALTER L. SMITH
University of North Carolina

CHAPMAN AND HALL

LONDON

First published 1961
by Methuen and Co. Ltd
Reprinted 1963, 1965, 1967, 1968
Reprint 1971 published by
Chapman and Hall Ltd
11 New Fetter Lane, London EC4P 4EE
First issued in limp binding 1971
Reprinted 1974

Printed in Great Britain by
Butler & Tanner Ltd
Frome and London

SBN 412 10930 1

Distributed in the U.S.A. by Halsted Press a division of John Wiley and Sons, Inc. New York

They also serve who only stand and wait

MILTON

Contents

Preface

Queues for service of one kind or another arise in many different fields of activity. In recent years a considerable amount of research has been conducted into the properties of simplified mathematical models of such queueing systems. Our objects in the present monograph are threefold. First we have tried, especially in Chapter I, to give an account of the general ideas that are useful in describing and thinking about queueing systems. Secondly, we have illustrated by examples some of the mathematical techniques that are useful for the study of these systems. Finally, we have given some explicit mathematical results which may be useful in practical investigations.

A recent bibliography gave some 600 papers on queueing and allied subjects. Clearly we cannot, in the modest limits of the present work, cover more than a small proportion of the huge amount of material available. However, some results that we have not had space to discuss in detail have been given in outline in the form of exercises.

We hope that this monograph will prove useful to students and workers in probability and statistics who want a short introduction to the problems of this special field. Primarily, however, we have written for the operational research worker who is concerned with practical investigations of queueing. We fully appreciate that nearly every practical problem has its own special complexities, but we believe that consideration of the general methods set out in this monograph may lead the way to an understanding of such special systems. It is also possible that an intelligent use of various results which we derive for particular simple systems will lead to useful approximate formulae for more complex systems.

The mathematical parts of the monograph require a knowledge only of the elements of probability theory and of advanced calculus. In particular we have assumed familiarity with the formal rules for manipulating expectations and generating functions. As far as possible we have adopted an applied mathematician's approach to the subject. For example, we have nowhere discussed such matters as the general conditions under which stationary probability distributions will exist. Such questions undoubtedly have their interest, but the applied worker is usually quite happy to rely on his intuition here, and it rarely misleads him in such matters!

Unfortunately it has not been possible to provide extensive numerical tables; we have aimed, however, at presenting results as far as possible in a form from which computation is straightforward.

After some consideration we decided not to burden the text with references to research papers and to other books. A beginner is apt to find such references irritating, and is depressed by the thought that perhaps he should read them all before he proceeds. Therefore, wherever possible, we have confined the references to Appendix I, where we give a brief discussion of collateral reading for each chapter, together with an indication of the main 'source' papers from which came the ideas presented in that chapter.

D. R. COX
Birkbeck College,
London

WALTER L. SMITH
University of North Carolina,
Chapel Hill, N.C.

CHAPTER I

Introduction

1.1 Some examples

We begin by discussing a number of examples of congestion. These will serve both to illustrate the wide range of applications covered by the theory and to introduce certain key ideas. The thing common to all the systems that we shall consider is a flow of *customers* requiring *service*, there being some restriction on the service that can be provided. For example, the customers may be aircraft requiring to take-off, the restriction on 'service' being that only one aircraft can be on the runway at a time. Or, the customers may be patients arriving at an out-patients' clinic to see a doctor. The restriction on service is again that only one customer can be served at a time, and, in fact, these two examples are both cases of what we shall call the *single-server* queue. An example of a multi-server queue is a queue for buying stamps at a British main Post Office, or for having goods checked at a supermarket. Here the restriction is that not more than, say, m customers can be served at a time.

Some other examples of systems where the number of customers that can be served at a time is restricted, are the following:

(*a*) articles pass along a conveyor belt and are to be packed into boxes;

(*b*) machines stop from time to time and require attention by an operative before restarting, the operative being able to attend to only one machine at a time;

(*c*) numerous problems connected with telephone exchanges;

(*d*) items of work (customers) come into, say, a testing department or a typing pool, the number of items that can be dealt with being limited by the number of workers in the department.

Some more applications are mentioned below and others will no doubt occur to the reader.

In the examples given so far, the restriction on service is that not more than a limited number of customers can be served at a time, and congestion arises because the unserved customers must queue up and await their turn for service. Sometimes, however, the restriction is that service is only available during limited periods; during these periods there may or may not be a limit on the number of customers that can be served at a time. For example, if the customers are pedestrians waiting to cross a busy road at an un-controlled crossing, service is available only when a sufficient gap develops in the traffic; when this occurs a large number of custom-ers may be able to cross (i.e. be served) simultaneously. If the customers are cars waiting to move at traffic lights, or waiting to move from a minor road into a major road, service is restricted both by the need to wait until a 'free period' occurs, and by the need to wait until previous customers in the queue have been served. That is, we have a mixture of the two types of restriction on service. Another example is a queue of people requiring a taxi; here, whether or not a particular customer can be served immedi-ately on arrival depends on whether or not taxis are free in the rank.

Another type of application, closely related to the single-server queue, concerns the size of stores. Suppose that the customers, in the previous discussion, are items placed from time to time in a store, but that not more than a limited number of items can be stored at once. The service of a customer is the withdrawal of an item from the store for use. We are interested, in this application, in the relation between the maximum content of the store and the chance that the store will be found empty, when a call for a fresh item is made.

1.2 The aims of an investigation of congestion

In all the examples in the previous section, congestion will occur from time to time if there is sufficient irregularity in the system. For example, in the single-server queue, suppose either that the customers arrive irregularly or that there is appreciable variation in the time taken to serve a customer, or both. Then from time to time more than one customer will be at the service point at the same time, all but one of them must queue up awaiting their turn for service, and congestion has occurred. This simple point illustrates an important general principle, namely that the congestion occurring in any system depends in an essential way on the irregularities in the system and not just on the average properties. When we come to mathematical analysis we nearly always specify irregularity in probabilistic terms and this is why the mathematics in the book is so heavily weighted towards probability theory.

Our practical aim in investigating a system with congestion is usually to improve the system by changing it in some way. For example, the rate of arrival of customers may be so high that large queues develop, resulting in a high waiting-time per customer, or the rate of arrival may be so low that the service facilities are unused for a large proportion of time. In either case a change in the system may be economically profitable. Or it may be that some radical reorganization of the system, such as the reduction in service-time by mechanization, may be under consideration. In any case, it is often very helpful to be able to predict what amount of congestion is likely to occur in the modified system. Not only is this useful in indicating which of several different modifications is likely to be the most rewarding for experimental study, but also, in certain applications, particularly in industry, it is impossible to test the modification experimentally before reaching a decision concerning its introduction. In this case the decision must be based either on guesswork, or on a theoretical prediction of what would be likely to occur if the modification were introduced.

For example, suppose that the customers are workers who from time to time need to grind their tools at a grinding machine. Production is lost while workers queue up awaiting their turn at the grinding machine. No theoretical analysis is needed to find the amount of this lost time; it can be measured along with the frequency with which grinding is required, the mean time taken to grind a tool, and other properties of the system. But suppose that the introduction of a second grinding machine is contemplated.

A rational decision about its introduction must depend, in part, on assessing the annual profit from the additional production obtained due to the reduction in queueing-time, and comparing this with the capital cost of the new machine. The additional production clearly cannot be measured experimentally prior to the decision about the new machine. Usually, however, we use theoretical results to predict what is likely to happen, and then check the prediction by small-scale experiments before embarking on the full practical application.

There are various ways of describing congestion, for example in terms of the queueing-time of individual customers, or in terms of the free and busy periods of the server. These will be discussed later. In order to predict one or more of these quantities we must specify the system sufficiently fully and this usually means giving:

(*a*) the arrival pattern. This means both the average rate of arrival of customers and the statistical pattern of the arrivals;

(*b*) the service mechanism. This means stating when service is available, how many customers can be served at a time, and how long service takes. Usually this last is specified by a statistical distribution of service-time;

(*c*) the queue-discipline. This means the method by which a customer is selected for service out of all those awaiting service. The simplest queue-discipline consists in serving customers in order of arrival, but there are many other possibilities.

The specification must also include an account of any interaction between the different parts of the system, such as a tendency for the server to work more quickly when there are a large number of customers waiting, and so on.

When the system has been described sufficiently fully, it becomes a mathematical problem to predict what the system will do. Of course, any description of the system in mathematical terms will inevitably oversimplify the practical situation, in virtue of the use we shall make of idealized concepts like a completely random series, and so on. However, the situation is in no way different from that in any other branch of applied mathematics, and simply means that we must apply the mathematical results critically.

The next step is to describe in somewhat more detail the types of arrival pattern, service mechanism and queue-discipline that are most used in the mathematical theory.

1.3 The arrival pattern

(i) *Regular arrivals*

The simplest arrival pattern physically, although not the easiest to deal with mathematically or the most common in applications, is the *regular* one in which customers arrive singly at equally spaced instants, a_1 units of time apart. The rate of arrival of customers is $\alpha = 1/a_1$ per unit time. A practical example where the arrivals are nearly regular is in certain conveyor belt systems. Another is when there is an appointment system closely adhered to.

(ii) *Completely random arrivals*

The simplest arrival pattern mathematically, and the most commonly useful one in applications, is when the arrivals are *completely random*. To define this formally, let there be a constant α, which will again represent the average rate of arrival of customers, such that for any short time interval $(t, t+\Delta t)$ the probability that no customers arrive is $1 - \alpha\Delta t + o(\Delta t)$, that one customer arrives is $\alpha\Delta t + o(\Delta t)$ and, therefore, that two or more customers arrive is $o(\Delta t)$, where the symbols $o(\Delta t)$ denote quantities that become negligible compared with Δt, as $\Delta t \to 0$. Further, what happens in this time interval is assumed to be statistically independent of the arrival or non-arrival of customers in any time interval not overlapping $(t, t+\Delta t)$.

To put this more vividly, consider Fig. 1.1. Arrivals of customers are denoted by points. The interval A, of length Δt, comes after a long gap with no arrivals; the interval B, likewise of length Δt, comes after a group of arrivals. In both cases the probability of an arrival in the interval Δt is the same, namely $\alpha\Delta t + o(\Delta t)$. Quite generally, the probability of an arrival in one interval is entirely unaffected by arrivals at other times. It follows that a completely

B

random series of arrivals is a very special form of arrivals; in fact whenever we use the term completely random we have in mind this particular form, and do not mean just a vaguely haphazard pattern.

The completely random series is likely to be a particularly good approximation when the customers are drawn from a very large pool of customers all behaving independently of one another. For example, the calls arriving at a telephone exchange over a fairly short time have a distribution closely approximated by a completely random series; if longer periods of time are considered, secular changes in the rate of calls will appear. Similar remarks apply to the arrival of aircraft at a busy aerodrome. Other examples of arrivals approximated by a completely random series are stop-

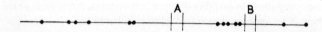

Fig. 1.1. Fundamental property of completely random arrivals. Two intervals A and B of equal length and therefore with equal probability for an arrival.

pages of a machine produced by mechanical breakdown or, in textile processing, by an end-breakage.

At this point it is convenient to derive, for future reference, some of the mathematical properties of the completely random series. First, the probability that two or more events occur in a short interval Δt, which has been assumed to be $o(\Delta t)$, is actually of order Δt^2; this can be obtained as a special case of a more general result to be derived in a moment. Secondly, it is useful to have the probability distribution of the number of customers arriving in a time period of fixed length t_0. To obtain this, divide the period t_0 into m intervals each of length Δt, $m\Delta t = t_0$. The probability that a single customer arrives in any one of these intervals is $\alpha\Delta t + o(\Delta t)$ and the probability that no customer arrives is $1 - \alpha\Delta t + o(\Delta t)$. Also the number of customers to arrive in different intervals are statistically independent, and the probability that two or more customers arrive in any interval is $o(\Delta t)$. Therefore, by the binomial probability law, the probability that just r customers arrive in the whole period is, in the limit

$$\lim_{\Delta t \to 0} \frac{m!}{r!(m-r)!} [\alpha \Delta t + o(\Delta t)]^r [1 - \alpha \Delta t + o(\Delta t)]^{m-r} \quad (1)$$

$$= \lim_{m \to \infty} \frac{m!}{r!(m-r)!} \alpha^r \frac{t_0^r}{m^r} \left(1 - \frac{\alpha t_0}{m}\right)^{m-r} \quad (2)$$

$$= \frac{(\alpha t_0)^r}{r!} \lim_{m \to \infty} \frac{m!}{m^r(m-r)!} \lim_{m \to \infty} \left(1 - \frac{\alpha t_0}{m}\right)^{m-r} \quad (3)$$

$$= \frac{(\alpha t_0)^r e^{-\alpha t_0}}{r!}. \quad (4)$$

Here in the passage from (1) to (2) the relation $m \Delta t = t_0$ has been used, and the terms in $o(\Delta t)$ neglected. In going from (3) to (4) we have used the result that, for fixed r,

$$\lim_{m \to \infty} \frac{m!}{m^r(m-r)!} = \lim_{m \to \infty} \left(1 - \frac{1}{m}\right) \dots \left(1 - \frac{r-1}{m}\right)$$

$$= 1.$$

Hence we have shown that if A is the number of arrivals in a period t_0, a random variable,

$$\varpi_r = \text{prob}(A = r) = \frac{e^{-\alpha t_0}(\alpha t_0)^r}{r!}, \quad (5)$$

and this, called the *Poisson distribution*, is one of the basic distributions of statistical theory. Fig. 1.2 gives an idea of the shape of the distribution for a few special values of αt_0.

We record the main properties of the Poisson distribution for reference. The mean and variance are both equal to αt_0; this is shown by summing the appropriate series, for example

$$EA = \sum_{r=0}^{\infty} r \varpi_r = \alpha t_0.$$

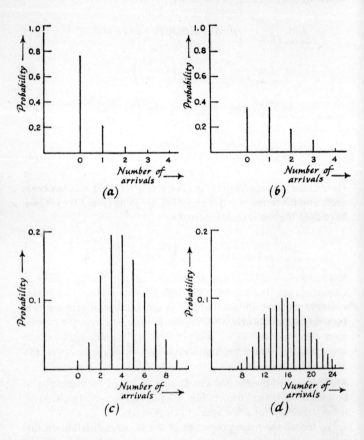

Fig. 1.2. Completely random arrivals. The Poisson distribution for the number of arrivals in a fixed time. (a) Mean, $\frac{1}{4}$; (b) Mean, 1; (c) Mean, 4; (d) Mean, 16.

Notice that α is the mean number of arrivals per unit time. Similarly

$$\text{var}(A) = EA^2 - [EA]^2$$

$$= \sum_{r=0}^{\infty} r^2 \varpi_r - \alpha^2 t_0^2$$

$$= \alpha t_0,$$

on some reduction. The probability generating function of the distribution is

$$E\zeta^A = \sum_{r=0}^{\infty} \zeta^r \varpi_r$$

$$= \exp[\alpha t_0(\zeta - 1)]. \tag{6}$$

The reader will find it a simple exercise to show that if t_0 is small, say equal to Δt,

$$\text{prob}(A \geqslant 2) = \sum_{r=2}^{\infty} \varpi_r = O(\Delta t^2). \tag{7}$$

The formulae (1)–(7) describe the distribution of the number of customers arriving in a period of time of fixed length t_0. It is sometimes useful instead to consider the distribution of the interval between successive arrivals. This can be obtained from the expression for $r = 0$ in (5), or alternatively by various more direct arguments, of which the following is particularly instructive. Fix an arbitrary time instant t_0, say, possibly, but not necessarily, at an arrival instant, and let

$$\Phi(x) = \text{prob}\{\text{no customer arrives after } t_0 \text{ and before } t_0 + x\}.$$

Then

$$\Phi(x + \Delta x) = \text{prob}\{\text{no customer arrives between the times } t_0 \text{ and}$$

$$t_0 + x, \text{ nor between the times } t_0 + x \text{ and } t_0 + x + \Delta x\}$$

$$= \Phi(x)\{1 - \alpha\Delta x + o(\Delta x)\},$$

by the product law of probability and the definition of the completely random series. Therefore

$$\frac{\Phi(x+\Delta x) - \Phi(x)}{\Delta x} = -\alpha\Phi(x) + o(1),$$

whence

$$\frac{d\Phi(x)}{dx} = -\alpha\Phi(x),$$

so that

$$\Phi(x) = a e^{-\alpha x},$$

where a is a constant of integration. Since $\Phi(0) = 1$, by definition, $a = 1$, and

$$\Phi(x) = e^{-\alpha x}. \tag{8}$$

Compare this with the expression for ϖ_0 from (5).

The probability density function (p.d.f.) of the interval between an arrival and the next subsequent arrival is therefore

$$\lim_{\Delta x \to 0} \frac{\text{prob}\{\text{interval lies between } x, x + \Delta x\}}{\Delta x}$$

$$= \lim_{\Delta x \to 0} \frac{\Phi(x) - \Phi(x + \Delta x)}{\Delta x}$$

$$= -\frac{d\Phi(x)}{dx}$$

$$= \alpha e^{-\alpha x}. \tag{9}$$

We call this the *exponential distribution*; notice that it is continuous whereas the Poisson distribution is discrete.

The following properties follow easily from (9) and are left as exercises for the reader.

(*a*) The mean interval is $1/\alpha$. This can also be seen from the general consideration that events occur at a rate α per unit time and that the mean interval between successive arrivals must equal the reciprocal of the rate of occurrence.

(b) The standard deviation of the interval is $1/\alpha$; the percentage coefficient of variation of the distribution is therefore 100.

(c) The ordinate of the frequency curve is greatest at $x = 0$ and decreases as the length of the interval, x, increases. Thus short intervals are relatively frequent and this suggests that a completely random series will show a considerable tendency to clump. The sections of a random series shown in Fig. 1.3 confirm this.

(d) The distribution (9) refers to the interval between successive arrivals, or to the interval from an arbitrary fixed instant to the next following arrival. That these two things should have the same distribution appears paradoxical at first sight, but it is in fact a

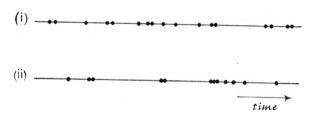

Fig. 1.3. *Sections of a completely random series of arrivals.*

natural consequence of our initial assumption that whether or not an arrival occurs in one interval is independent of what happens in other intervals. Hence whether the starting point corresponds to an arrival or not can have no effect on the subsequent occurrences, and in particular on the length of time elapsing before the next arrival.

(e) In view of the property discussed at the end of (d), the interval from say the arrival of the nth customer to that of the $(n+1)^{st}$, and the interval from the $(n+1)^{st}$ to the $(n+2)^{nd}$ are statistically independent. Quite generally, the intervals between the arrivals of successive customers are mutually independent random variables, all having the probability distribution (9). This may be used as a definition of a completely random process.

A final property of the completely random series that is some-times required is the distribution of the interval from a fixed point, e.g. an arrival, to the kth following arrival. The case just discussed is $k = 1$. The distribution required is the sum of k independent random variables each having the distribution (9). The Laplace transform of (9) is

$$\int_0^\infty e^{-sx} \alpha e^{-\alpha x} dx = \frac{\alpha}{\alpha + s} \qquad (10)$$

and so the Laplace transform of the required distribution is $\alpha^k/(\alpha + s)^k$, which is recognized as belonging to the probability density function

$$\frac{\alpha(\alpha x)^{k-1} e^{-\alpha x}}{(k-1)!} \quad (x \geqslant 0). \qquad (11)$$

We have used here the well-known property of Laplace transforms that the Laplace transform of a convolution of two functions is simply the product of the transforms of the two functions. The distribution (11) is graphed for a few values of k in Fig. 1.4.

To sum up the more important properties of the completely random series: the number of arrivals in a fixed time t_0 follows the Poisson distribution (5); the intervals between successive arrivals are independently distributed in the exponential distribution; the interval from a given point up to the kth following arrival has the distribution (11).

(iii) *General independent arrivals*

The next type of arrival pattern is a natural generalization mathematically of the previous two. Suppose that there is a proba-bility distribution, defined by a distribution function $A(x)$, say, such that the intervals between the arrivals of successive customers are random variables, independently distributed in the given form. If $A(x)$ corresponds to the discrete distribution concentrated at a, we have case (i), regular arrivals. If $A(x)$ corresponds to the exponential distribution (9)

$$A(x) = 1 - e^{-\alpha x},$$

we have case (ii), completely random arrivals. If in a particular

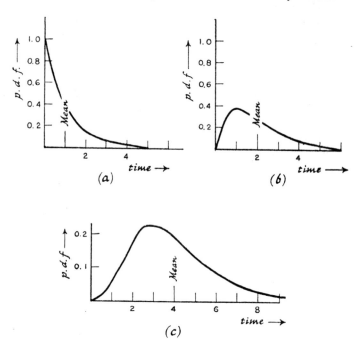

Fig. 1.4. *Completely random arrivals at unit rate. The distribution of the interval up to the kth arrival. (a) k = 1; (b) k = 2; (c) k = 4.*

mathematical analysis we take the inter-arrival intervals to be independent, with a general distribution function $A(x)$, we shall say that the arrivals have the *general independent* form.

The condition that the different intervals are statistically independent means, for example, that if an interval between two arrivals is by chance particularly long or particularly short, the distribution of the next interval is unaffected.

One example of an independent input is an m-server queue in which a completely random stream of customers is assigned to the different servers as follows. Customers are numbered in order of arrival, and numbers $1, m+1, 2m+1, \ldots$ go to server 1, numbers $2, m+2, 2m+2, \ldots$ go to server 2, and so on. Then the arrival pattern for any particular server can be seen to have the independent form with a distribution of intervals obtained by replacing k in (11) by m.

Another example is when customers arrive in one queue immediately after they have been served in a preceding queue, in which the server is never free. The arrival pattern for the second queue is then the independent form with the distribution function $A(x)$ equal to the distribution function of service-time in the first queue.

It can be shown that the variance of the number of customers arriving in a long time t is asymptotically $C^2 t$ times the mean inter-arrival interval, where C is the coefficient of variation of the distribution $A(x)$. This generalizes the result in (ii) that the variance is exactly equal to the mean for the completely random series.

(iv) *Regular arrivals with unpunctuality*

Suppose that customers have appointments to arrive at equally spaced intervals a, but are unpunctual. More precisely assume that the nth customer, scheduled to arrive at time na, actually arrives at $na + \epsilon_n$, where $\epsilon_1, \epsilon_2, \ldots$ are independently distributed random variables, whose distribution characterizes the amount and form of the unpunctuality. If the ϵ_i's have mean zero, there is no tendency to be on the average late or early for the appointment.

In general, if the dispersion of the ϵ's is small compared with a, the effect of unpunctuality is unimportant. If the dispersion of the ϵ's is large compared with a, the series is equivalent to a completely random one.

(v) *Aggregated arrivals*

Up to now we have considered situations in which customers arrive singly. It sometimes happens that customers arrive in groups of varying sizes. A fairly general set-up of this type can be described as follows. There are a series of *arrival instants* distributed in one of the ways we have discussed above. There is also a group size distribution, a probability distribution over the non-negative integers, such that the numbers of customers at the different arrival instants are independent random variables with this distribution.

For example, the arrival instants might be regularly spaced and the group size distribution might have equal probabilities of $\frac{1}{3}$ at the values 0, 1, 2.

(vi) *Complex deterministic arrivals*

In the cases discussed up to now, any irregularity in the arrivals is statistical in character and is described mathematically in terms of random variables. Occasionally, however, irregularity is produced by a complex recurring pattern. Thus, a semi-automatic machine might require unloading every 15 seconds, reloading with fresh raw material every 2 minutes and general adjustment every 6 minutes. If an operative has several such machines under her charge, the pattern of arrivals, i.e. the series of instants at which work is required from the operative, has a complicated but completely deducible structure. Very often such patterns can be treated as completely random, although no general statement can be made about this.

(vii) *Discrete-time arrivals*

Sometimes, usually for mathematical convenience, we represent an irregular pattern of arrivals by a series in which arrivals can occur only at a discrete set of time instants 0, h, $2h$, . . ., say. For example if we wanted to approximate to a completely random series in this way, we let $p = \alpha h$ and have a probability p of an arrival taking place at a particular instant, independently for different instants. If $p \ll 1$, i.e. if $h \ll 1/\alpha$, we get a close approximation to the completely random series in continuous time. Similar approximations may be constructed for the other arrival patterns given above.

From a mathematical point of view these schemes are closely related to aggregated arrivals, (v), in which the arrival instants are equally spaced and the number of arrivals per instant is 0 or 1. However, it seems worthwhile to keep the present application distinct.

(viii) *Non-stationary arrival patterns*

All the arrival patterns described above are what is known as *stationary*, i.e. the probability structure of the process does not vary with time. Sometimes, however, we are interested in non-stationary arrivals, and the following are a few examples.

If the customers are telephone calls put into an exchange, the arrival pattern is completely random at a rate that varies smoothly with the time of day. If customers are patients with a particular type of disease requiring a bed in a hospital, the arrival pattern is likely to be completely random at a rate depending on the time of year, and possibly also on the day of the week. In certain 'rush-hour' problems, we can reasonably approximate to the arrival pattern by taking, say, a completely random series in which the rate of arrival is to begin with α_0, suddenly changes to α_1, and, after a certain time, returns either to α_0 or to some other value. Again, in certain industrial processes, failures occur completely randomly at a rate which varies somewhat from one consignment of raw material to another.

All these possibilities are covered by taking a stationary pattern, and allowing the parameters in it, for example the arrival rates, to vary either smoothly or discontinuously.

(ix) *Arrivals correlated with other aspects of the system*

Sometimes the rate of arrival of customers is correlated with other properties of the system, usually with the number of customers awaiting service. The possibilities here are many. For example, in a shop queue, customers may be deterred or attracted by the presence of a long queue of customers. Or in certain industrial problems, the arrival of new customers may be cut off completely as soon as the number of unserved customers reaches a critical value. Or there may be a certain chance that a customer will leave the system unserved if his service is not begun sufficiently quickly.

A further very important case is when there is only a finite number, n, of possible customers. An example is when the customers are machines, each operative having n similar machines in her charge. Then suppose that any one particular machine that is running at some instant has a chance $\alpha \Delta t + o(\Delta t)$ of stopping, i.e. arriving for service, in the next interval Δt. Then if r machines are stopped at any time, i.e. are being served or awaiting service, the chance of a further arrival is $(n-r)\alpha \Delta t + o(\Delta t)$. In other words we have a completely random series of arrivals in which the parameter varies depending on the state of the system.

Most of the other examples can be dealt with in a similar way. We take one of the basic patterns, usually the completely random series, and allow the parameters in it to be functions either of the total number of customers awaiting service, or sometimes, of the *cumulated service-time*. This last quantity is, for a single-server queue, defined, at any instant, to be equal to the sum of the service-times of the unserved customers and the remaining time necessary to complete the service of the customer in course of being served. If customers are served in order of arrival, the cumulated service time is equal to the total time that a fresh customer must wait before his service begins. If the cumulated service-time is known to potential customers, it is probable that this, rather than just the number of customers, will determine their decision whether or not to enter the system.

(x) *Arrivals in a continuous flow*

In the previous discussion, and in the great majority of applications, customers may be treated as discrete individuals, whose arrival takes place at well-defined instants in time. Occasionally, however, the customers are to be treated as a continuous flow. An example would be a storage system in which, say, a fluid arrives in the store in a continuous stream, at a constant or variable rate. The process of removal from the store, i.e. of service, might be continuous in another flow, or in discrete batches. One such example would be the study of the size and number of gasholders advisable in a certain plant.

The arrival in these cases is described by a continuous time-series, obtained by plotting the rate of arrival against time.

The above discussion shows that there is a wide variety of arrival patterns that can arise in applications and, of course, there are many other possibilities from a formal mathematical point of view. In fact, the completely random and regular patterns are the most commonly used in applied mathematical work and it is only for these that mathematical solutions of any generality can be obtained; other arrival patterns usually require special investigation.

1.4 The service mechanism

In the previous section we discussed the arrival of customers. We now turn to the servicing operation. There are three aspects of this that need description. First, there is the length of time taken to serve an individual customer, the *service-time*. In the great majority of cases we assume that the service-times of different customers are independent random variables, all with the same probability distribution, to be called the *service-time distribution*. In more complicated cases the customers may be of several types, each with its own distribution of service-time.

The second aspect of service is that of the *capacity* of the system. This is defined as the maximum number of customers that can be served at any one time. For example, in the single-server queue discussed above, the capacity is one, for the *m*-server queue, the capacity is *m*. In transport problems, the capacity is the maximum number of customers that can be carried per journey.

The third property of service is its *availability*. To describe this we must state both when service facilities are available and also any restrictions which reduce the number of customers that can be served together, below the full capacity of the system. For example, if we consider a 'bus queue, we specify the series of instants at which 'buses arrive and the probability distribution of the number of free seats per arrival. In the simple single-server queue and the m-server queue as described above, the servers are always present to deal with customers; we call these systems with *complete availability*. The majority of published mathematical work concerns systems with complete availability, but in practice it is often possible for one or more of the servers to leave the service point from time to time. To make a mathematical analysis of the consequences of this incomplete availability we must specify, usually in probability terms, the frequency and duration of the periods of absence. There are a great many possibilities here.

We now discuss the three elements of service in a little more detail.

(i) *The distribution of service-time*

There are some calculations connected with queueing processes that can be carried through with a general distribution of service-time, assuming only that this distribution is constant in time and that the service-times of different customers are statistically independent. More usually, particularly in fairly complicated problems, it is helpful to assume that the distribution is of some special type of which the following are the most common.

(a) *Constant service-time.* The service-time may be assumed to be constant. This is always an idealization but, particularly in problems with very irregular arrival patterns, it often gives adequate answers.

(b) *Exponential service-time.* A great mathematical simplification results if we can reasonably represent the p.d.f. of service-time by the exponential curve

$$\sigma e^{-\sigma x} \quad (x \geq 0). \tag{12}$$

This has its mean and its standard deviation equal to $1/\sigma$. Now if the random variable X is a service-time,

$$\text{prob}(X \geq x_0) = \int_{x_0}^{\infty} \sigma e^{-\sigma x} dx = e^{-\sigma x_0}. \tag{13}$$

From this result we can calculate the probability that the service of a customer is completed in an interval Δx, given that the service has been in progress for a time x_0. This is the conditional probability

$$\text{prob}(x_0 \leqslant X \leqslant x_0 + \Delta x \,|\, X \geqslant x_0)$$

$$= \frac{\text{prob}(x_0 \leqslant X \leqslant x_0 + \Delta x)}{\text{prob}(X \geqslant x_0)}$$

$$\sim \frac{\sigma e^{-\sigma x_0} \Delta x_0}{e^{-\sigma x_0}}$$

$$= \sigma \Delta x_0. \tag{14}$$

In other words the probability that service is completed in a small element of time is constant, independently of how long service has been in progress. Thus service can be treated as if it were a completely random operation.

This result should be compared with the discussion of completely random arrivals leading to (9). There we started from a definition analogous to (14) and deduced the exponential distribution. Here we start from the exponential distribution and deduce (14).

The exponential distribution is a good approximation to the distribution of the duration of telephone calls. It is likely to be a reasonable thing to consider when there are a large number of customers requiring fairly short service and a smaller number of customers requiring longer service.

(c) Γ *type and special Erlangian service-time*. The two distributions considered so far have coefficients of variation of 0 and 100 per cent respectively. It is very common in applications to have a unimodal frequency distribution with a coefficient of variation intermediate between these values, and so it is useful to have a simple flexible mathematical formula that can be used to approximate to practical distributions of this type. Such a formula is that of the Γ type frequency curve, namely

$$\frac{1}{\Gamma(k)}\frac{k}{b_1}\left(\frac{kx}{b_1}\right)^{k-1} e^{-kx/b_1} \quad (x \geqslant 0). \tag{15}$$

This has mean b_1 and coefficient of variation $100/\sqrt{k}$.

The special case $k = 1$ is the exponential curve (12) with $b_1 = 1/\sigma$, and the limit as k tends to infinity corresponds to a constant service-time of b_1. Equation (15) represents a frequency curve for all positive values of k and therefore a curve (15) can be found with any assigned mean and standard deviation. However, from a mathematical point of view, particular interest attaches to the case when k is an integer. Since $\Gamma(k)$ is then equal to $(k-1)!$, equation (15) is then of exactly the same form as (11), representing the distribution of the interval between k events in a completely random series. Fig. 1.4 illustrates the form of the distributions for a few values of k.

We shall call the distribution (15) with k an integer a *special Erlangian* distribution. In view of the remark at the end of the last paragraph, we can, if the service-time distribution is of this form, regard the servicing operation in the following way. There are k stages, not necessarily having physical significance. The lengths of time to complete the different stages are independent random variables all with the distribution $(k/b_1)e^{-kx/b_1}$. When the first stage is completed, the second stage is begun, and so on, the servicing operation being completed at the end of the kth stage. The total service-time then has the distribution of the sum of k independent random variables, each with the above exponential distribution, and hence is distributed in the form (15). This link between the special Erlangian distribution (15) and the exponential distribution is a very useful one, and is developed further in section 5.2.

(d) *General Erlangian service-time.* The Laplace transform of the special Erlangian distribution (15) is $(1 + sb_1/k)^{-k}$, and is the reciprocal of a special polynomial of degree k. Some of the

C

simplification resulting from the use of (15) is retained for any distribution with a rational Laplace transform, i.e. with a Laplace transform $P(s)/Q(s)$, where P and Q are polynomials in s. We shall call the class of distributions of this type the class of *general Erlangian* distributions.

If the polynomials $P(s)$ and $Q(s)$ have no common factors, it is necessary, in order that $P(s)/Q(s)$ correspond to a distribution of a non-negative random variable, that the zeros of $Q(s)$ have negative real parts, and that the degree of $P(s)$ should not exceed that of $Q(s)$. We can approximate arbitrarily closely to any distribution of service-time by taking k sufficiently large, but in practice Erlangian distributions with large k and general Erlangian distributions with complicated $P(s)$ and $Q(s)$ are difficult to handle. These distributions are of most value when either k is small or when the polynomials $P(s)$ and $Q(s)$ have some simple form.

The special distributions (a) to (d), or simple mixtures of them, are the ones most commonly used to represent the distribution of service-time, when the service-times of different customers can be represented as independent identically distributed random variables. We now discuss ways in which this last, independence, assumption is invalid.

(e) *Non-stationary service-time*. The distribution of service-time may change with time due, for example, to fatigue of the server.

(f) *Service-time correlated with other aspects of the system*. In section 1.3 (ix) we considered briefly arrival patterns that are correlated with the amount of congestion in the system. Similar things can happen with the service-time. For example, the server may work faster, or slower, when the number of customers awaiting service, or the cumulated service-time, is high. Variation of this type must not be confused with random variation of service-time.

(g) *Customers of several types*. It may happen that there are several types of customers, each type having its characteristic distribution of service-time. If the queue-discipline takes no account of the differences between the types of customer, we may be able to ignore the differences between the customers and take the distribution of service-times to be the pooled distribution obtained by combining the separate distributions in appropriate proportions. In general, however, the distinction between the distributions will be important.

(ii) *Service capacity*

The next aspect of service for discussion is the capacity, the maximum number of customers that can be served at a time. The case most commonly discussed in mathematical work is when the capacity is one. Practical examples of such systems were given at the beginning of the chapter. Systems with capacity m, for a general integer m, were also described there. Another example is in the study of hospital appointment systems, in which the aspect of interest is not the length of time spent queueing after arrival at the hospital, but instead the number of days elapsing from first requesting an appointment until the day of the appointment. In this case the capacity of the system is the maximum number of patients that can be dealt with in one clinic. Systems of unlimited capacity can arise in mathematical work, when we assume that the only restriction on service is in its availability; for example, in a study of pedestrian crossings, it might be assumed that as soon as a sufficient gap develops in the traffic, an unlimited number of pedestrians can cross.

(iii) *Service availability*

We have now discussed the length of time that service takes and the capacity of the service system and so turn to the availability of service. There are very many possibilities here and it does not seem practicable to cover them in any very systematic way. In systems with capacity one we have to specify the frequency and duration of the server's absences; these may be independent of the congestion of the system, for example, 6 minutes' absence at the end of every hour, or may be dependent on the congestion, for example, absence 10 per cent of the time, periods of absence being begun only when there are no customers for service. In either type of situation we must, for mathematical purposes, specify the statistical nature of the non-availability of service fairly precisely before calculations can be made; this sometimes causes difficulty in applications in that, say, the time distribution of servers' absence is determined largely by personal whim and cannot be represented in a rigid mathematical scheme.

In systems with capacity more than one, we have in effect to specify the distribution of capacity in time. In some cases, the servers will either be all absent or all present, and then we describe the periods of absence in ways similar to those used for the single capacity problem. In other cases the possibilities are much more numerous; thus in the 'bus queue problem we need to give the joint distribution of the instants of arrival and of the number of free seats.

In certain problems the service-time is assumed negligible, and the system to be of unlimited capacity; the whole mathematical problem is then one of calculating the distribution of availability.

1.5 The queue-discipline

The final element in the description of a congestion system is the queue-discipline, which specifies how customers are to be selected for service from the pool of customers who have arrived at the queueing point. We shall consider queue-discipline separately for single capacity and for multiple capacity systems.

(i) *Single capacity systems*

When the service of a customer is completed, or when the service becomes available, one customer must be selected for service. This may be done on the basis of the order of arrival of customers, or on the basis of some prior classification of the customers.

In the first case, the commonest procedure is to serve customers in order of arrival. Other possibilities are to select a customer at random with respect to order of arrival, or to take the last customer to arrive rather than the first. Effectively random selection occurs in many telephone systems and in queueing for taxis in New York. 'Last come, first served' is the rule in some industrial problems in which unserved articles are stored in a container, the last article placed in being the most accessible.

In the second case, when the rule for selection does not depend solely on order of arrival, there are two main possibilities. The queue-discipline may depend on some prior numbering of the customers, for example in terms of a nominal appointment time. This numbering would usually be closely related to, but not necessarily the same as, the order of arrival. The other possibility is that the customers are divided into types, either with different distributions of service-time, or such that the loss due to delaying a customer unit time is different for different types of customer. Customers are then selected for service giving customers of Type 1 priority, and so on. In extreme cases the server may stop the service of a customer of lower priority in order to deal with a customer of high priority; this is called preemptive priority.

(ii) *Multi-capacity systems*

It may happen that certain servers specialize in the service of customers of certain types; this is advantageous if specialization enables the service-times to be reduced appreciably. In such cases we can usually treat the separate groups of servers and customers in isolation.

When any customer can be served equally well by any server, there are three main queue-disciplines, in which the customers are assigned to servers in strict rotation, or in which each customer decides on arrival which queue to join, or in which the customers form into a single queue, a customer moving forward for service as soon as a server becomes free. The first is simple mathematically, but inefficient practically and not often realistic. The second and third systems are those that arise most commonly in applications, but are difficult to deal with mathematically. With the second system, which is effectively that used in many banks and Post Offices, some interchange from one server's queue to another may be permissible if a long queue occurs at one point with a server free elsewhere. This makes the second queue-discipline about equivalent to the third. If no interchange is permissible the second system is less efficient than the third, although it may be more convenient practically.

1.6 The measurement of congestion

We now consider briefly some of the properties of a queueing system that may be of practical interest and which we may wish to calculate mathematically. There are three main ones, namely

(*a*) the mean and distribution of the length of time for which a customer has to queue for service;

(*b*) the mean and distribution of the number of customers in the system at any instant;

(*c*) the mean and distribution of the length of the server's busy periods. For a single-server queue a busy-period is defined to begin when a customer arrives to find the server free and to last until the next instant at which the server is free.

These three properties of the queueing system are related in a general way, in that all three mean values tend to increase as a system becomes more congested, but we shall, in any particular practical application, normally not be interested in all three quantities.

We define a customer's *queueing-time* to be the time from his entering the system to the instant at which his service begins, whereas his *waiting-time* is the time from entering the system to the instant at which his service is complete. Thus waiting-time equals queueing-time plus service-time. In many systems queueing-time and service-time are statistically independent and it is then easy to obtain properties of waiting-time from those of queueing-time, and vice-versa. In this book we shall therefore give results in terms of whichever of the two is the more convenient, although in a practical application there may well be reason for preferring to work with one property rather than the other.

Customer's queueing-time or waiting-time are of direct interest when there is an economic loss if a customer is kept queueing. If the loss per unit delay is constant, only the mean queueing- or waiting-times need be considered. Instances where the distribution and not just the mean are of interest arise when a customer may leave the system if delayed a long time, or when there is a penalty if a customer is delayed longer than some critical time.

For example if the customers are items of material that have passed through one stage of processing and are waiting for a second stage, it may happen that small delays are of no consequence, but that long delays will spoil the final product. In such a situation we shall be interested in the distribution of queueing-time.

The study of the number of individuals awaiting service is of particular interest when it is difficult to house customers awaiting service. The third variable, the length of busy periods, is probably the one most rarely required in practice but is needed sometimes, for example if attention is concentrated on the service mechanism and it is desired to arrange that long busy periods occurs infrequently.

In special problems other quantities may be required. Thus in a problem in which customers arriving during a period of high congestion are 'lost', one would be particularly concerned to calculate properties of the number of 'lost' customers. In general, in a practical application, one should aim to characterize congestion by the quantity with the most direct practical interpretation, and usually this means the quantity most directly related to the economics of the problem.

1.7 The investigation of queueing systems

In this monograph we have deliberately confined ourselves to the *mathematical* aspects of queueing phenomena. Even so, it will be appreciated that the number of combinations of arrival pattern, service mechanism and queue-discipline is large, and that it is not possible in a monograph of the present length to do more than outline some of the more important mathematical techniques that have been used on these problems, and to give explicit results for a few commonly occurring situations. Indeed, mathematical solutions are at the present time available only for the relatively more simple situations. If numerical results are needed for a complex system these have to be obtained by sampling experiments, a procedure that has lost some of its tedious character now that high-speed computing facilities are widely available.

It will be useful to put the mathematical side of the subject in its correct perspective by reviewing the steps that may enter into a practical investigation of a queueing problem. These are

(i) a preliminary survey;
(ii) a rough assessment of possible modifications of the system;
(iii) the obtaining of further data;
(iv) the detailed study of the effects of modification of the system;
(v) the formulation of a practical recommendation;
(vi) a small-scale trial of the modification;
(vii) full-scale practical action.

Of course, this list is intended as a general guide only, and in particular applications some stages may be omitted or perhaps some parts may need repetition.

The steps (i)–(vii) are discussed briefly below.

(i) *Preliminary survey*

The main steps here are likely to be

(a) in a complex system, the construction of a flow diagram for customers and a listing of the points at which there is a restriction on service;

(b) the approximate measurement of arrival rates, service-times and queueing-times at the main points of congestion.

This will give a general idea of the main points at which excessive congestion occurs, or at which there are service facilities which are idle for a high proportion of the time.

(ii) *Rough assessment of possible modifications*

There will usually in theory be many ways in which the system could be modified; a list of some such changes is given in the next section. At this stage those types of change are determined that are both practicable and reasonably likely to lead to an increase in efficiency, i.e. the general lines of the detailed investigations are settled.

(iii) *More detailed data*

In this stage the statistical properties of the input and service-time are estimated in sufficient detail to help in choosing an appropriate model and to enable the parameters in the model to be given numerical values. Thus in a simple single-server queue in which arrivals are thought to be random, it might be enough to obtain data to estimate the arrival rate, to provide a rough check on the randomness of arrivals, and to estimate the frequency distribution of service-times. It commonly happens that arrivals are random over limited periods but that the arrival rate changes systematically, say over a day; in such cases the systematic changes must, of course, be determined. In more complicated systems, with the possibility that the arrival rate and service-times depend on the number of customers in the system, or in which there are systems of priority, it will probably be necessary to make a detailed record of the behaviour of the system. In all cases where reliable quantitative results are required the use of correct sampling methods is essential.

In addition to the quantities describing input, service-time, etc., it will usually be desirable to measure the congestion by recording queueing-times, or the distribution of the number of customers in the queue; these can be used to check the applicability of any formulae that is proposed to use in stage (iv).

(iv) *The effects of modification*

In this stage we try to predict the effects of the type of modification under consideration. This can be done in three ways, by direct experiment, usually by a factorial design, by theoretical calculation, or by a simulation study. The first of these will often be ruled out; for example if the object of the investigation is to decide whether an expensive modification of a system is justified, direct experiment will usually be excluded. In both a theoretical and in a simulation study, it will be assumed that some parameters of the system, for example arrival rates, remain unchanged after modification, and hence predictions of relevant properties of the modified system are obtained.

As suggested in (iii) it will often be a useful check on the method to see whether theoretical queueing-times, etc., for the original system agree reasonably with observation.

Quite often a combination of simulation and mathematical analysis is effective, the simulation being used, for example, to check on simplifying assumptions made in the mathematical analysis.

(v) *Formulation of practical recommendation*

This will involve choosing from among the practicable modifications examined in (iv) the one that satisfies some optimality criterion. The criterion may either be expressed solely in terms of cost, or may be that a cost is to be minimized subject to some such condition as that the probability that a customer's waiting-time exceeds, say w_0, is not to exceed, say α_0. In general, optimality criteria, of whatever type, have to be examined critically.

(vi) *Small-scale trial*

Wherever possible a small-scale trial of the modification proposed in (v) is very desirable. This has the object of detecting any practical snags that have been overlooked in the theoretical analysis, and of checking that theoretical predictions made in (iv) are reasonably accurate. It is advisable to make sufficiently detailed observations to be able to interpret any discrepancy between say observed queueing times and those predicted in (iv). A serious discrepancy between theoretical and observed behaviour may be due either to an inadequate theoretical analysis or simulation study, in which perhaps some important effect has been ignored, or to a change in the parameters of the system that were assumed in the theoretical analysis to remain constant. Thus it might happen for reasons connected or unconnected with the modification, that the arrival rate during the small-scale trial is seriously different from that in the initial period, (i) and (iii).

(vii) *Full-scale practical action*

The final step is the full-scale introduction of the new system. Observation of the modified system from time to time may well be advisable.

1.8 The modification of queueing systems

In many practical applications the types of change in the system that are practicable and likely to bring practical advantages will be fairly clear. Nevertheless it may sometimes be useful to have a list of the more common modifications and this we now give.

In a complex system in which each individual passes in turn through several queueing points, it may be possible to reorganize the flow pattern. If, however, we are concerned with queueing at a single point, modifications may be made to the arrivals, to the service mechanism, and to the queue-discipline.

(i) *Modifications to arrivals*
We may

(*a*) modify the total mean arrival rate by, for example, excluding some customers from service;

(*b*) control the arrival times of individual customers by an appointment system designed usually to produce regular arrivals;

(*c*) even out systematic variations in arrival rate by, for example, trying to ensure a more uniform flow of customers over the day, without controlling arrival times of individual customers;

(*d*) arrange that customers are encouraged to join or discouraged from joining the queue, depending on the number of customers currently in the queue.

(ii) *Modification to service mechanism*
We may

(*a*) decrease the mean service-time;

(*b*) reduce the coefficient of variation of service-time;

(*c*) arrange that the service-times are reduced during periods of more than average congestion;

(*d*) change the capacity of the system by providing more servers;

(*e*) arrange that the capacity can be increased temporarily either when congestion is observed to be high, or when more than the average number of customers are expected to arrive;

(*f*) increase the server availability, either on the average, or by arranging that service is more likely to be available when customers are present.

(iii) *Modification of queue discipline*

We may

(*a*) give priority, preemptive or non-preemptive, to 'important' customers, i.e. to ones for whom the cost per unit waiting-time is high;

(*b*) give priority to customers whose service-times are expected to be short;

(*c*) introduce into a system in which the queue-discipline is not 'first come, first served' some device to ensure that the probability of very long queueing-times is reduced;

(*d*) in a multi-server queue, modify the arrangement by which customers are allocated to a particular server. There are several ways of doing this.

CHAPTER II

Some Simple Queues with Random Arrivals

2.1 Introduction

In this chapter we illustrate some of the mathematical methods that can be used in connexion with queueing problems. We consider for the most part a single-server system with complete availability.

As noted in section 1.6 there are three main properties of such a system that may interest us, namely the queueing-time, the number of customers in the system, and the server's busy periods. Before investigating these for some simple queueing systems, it is necessary to explain the important idea of an equilibrium probability distribution, and we do this in the next section in terms of a very simple example.

2.2 Statistical equilibrium

Suppose that we are studying the behaviour of a system which at any instant is in one of two possible states, A and B. Suppose that if the system is in state A at time t then the probability that it may switch to state B in the time interval $(t, t + \delta t)$ is, to the first order, $\alpha \delta t$ independent of all behaviour of the system prior to the time t. Similarly, for transitions from state B to state A we introduce a probability differential $\beta \delta t$.

An interpretation of this probability model as a queueing problem is that state A may represent the system with no customers present, state B the system with one customer being served. It is assumed that any customers that arrive while a customer is being served, state B, leave immediately without being served and do not return. This is, for example, an approximate representation of a single telephone receiver, taking only incoming calls, it being assumed that a caller finding the line occupied does not call again. Arrivals are assumed to be at random with parameter α and service-time to be exponentially distributed with parameter β.

Write $p_A(t)$, $p_B(t)$ for the probabilities that at time t the system is in states A, B respectively. Then elementary probability considerations lead us to the equations

$$\left.\begin{aligned} p_A(t+\delta t) &= (1-\alpha\delta t)\,p_A(t) + \beta\delta t\,p_B(t), \\ p_B(t+\delta t) &= \alpha\delta t\,p_A(t) + (1-\beta\delta t)\,p_B(t). \end{aligned}\right\} \tag{1}$$

In fact the probability that state A is occupied at time $t+\delta t$ is the sum of the probabilities that

(a) state A is occupied at time t and no transition occurs in the interval $(t, t+\delta t)$;

(b) state B is occupied at time t and a transition from B to A occurs in $(t, t+\delta t)$.

The probability of two or more transitions in the time interval is, by the product law of probability, of the order of $(\delta t)^2$ and can be neglected in forming the equations. Now the probability of event (a) is equal to $p_A(t)$ times the conditional probability, given that state A is occupied at time t, that no transition occurs, and this conditional probability is $1-\alpha\delta t$. Hence the probability of event (a) is $p_A(t)(1-\alpha\delta t)$. The probability of event (b) is calculated by a similar argument.

The type of argument just outlined is very important. For it to be used it is essential that the transition probabilities for the interval $(t, t+\delta t)$ shall depend only on the state occupied at t and shall be independent of the previous history of the system.

By passing to the limit in equations (1), we have that $p_A(t)$ and $p_B(t)$ satisfy the pair of differential equations

$$\left.\begin{aligned} \frac{d}{dt}\,p_A(t) &= -\alpha p_A(t) + \beta p_B(t), \\[2mm] \frac{d}{dt}\,p_B(t) &= \alpha p_A(t) - \beta p_B(t). \end{aligned}\right\} \tag{2}$$

Since we must have that $p_A(t) + p_B(t) = 1$, for all t, the differential equations (2) are easy to solve. We find

$$
\left.
\begin{aligned}
p_A(t) &= \frac{\beta}{\alpha+\beta}[1 - e^{-(\alpha+\beta)t}] + p_A(0)e^{-(\alpha+\beta)t}, \\
p_B(t) &= \frac{\alpha}{\alpha+\beta}[1 - e^{-(\alpha+\beta)t}] + p_B(0)e^{-(\alpha+\beta)t},
\end{aligned}
\right\}
\tag{3}
$$

where $p_A(0)$, $p_B(0)$ are the probabilities of the states A, B at the time origin.

We call equations (2) the differential equations of the process. In our later studies of congestion processes we shall discover that it is frequently possible to write down the appropriate differential equations of the process. It will often happen, however, that we are unable to present their solutions in a compact expression analogous to (3). Nevertheless this is not quite such a drawback as we might suppose, for reasons which will shortly emerge.

Notice that as $t \to \infty$ the probabilities in (3) approach the limiting values

$$
\left.
\begin{aligned}
p_A(\infty) &\equiv \lim_{t \to \infty} p_A(t) = \frac{\beta}{\alpha+\beta}, \\
p_B(\infty) &\equiv \lim_{t \to \infty} p_B(t) = \frac{\alpha}{\alpha+\beta}.
\end{aligned}
\right\}
\tag{4}
$$

Note that $p_A(\infty) + p_B(\infty) = 1$, so that these limiting values do provide a probability distribution. We call this the *equilibrium probability distribution* of the process. If $t(\alpha+\beta) \gg 1$, equation (3) shows that little error would be committed if the probabilities $p_A(t), p_B(t)$ were assigned their limiting values $p_A(\infty), p_B(\infty)$. Thus, if we were asked to wager on the state of the system at any given time, and we were not told exactly when the system had been 'switched-on', we would be fairly safe in basing our wager on the equilibrium probability distribution, provided we could be sure that the 'switching-on' took place some considerable time prior to the wager. Moreover it would, under such circumstances, be of little value to us to be told in what state the system found itself at the 'switching-on'; for it is evident from (3) and (4) that the equilibrium probability distribution is in no way influenced by the values assigned to $p_A(0)$ or $p_B(0)$.

When we can assume that the equilibrium probability distribution exists, its form can be found quite rapidly from the specification of the stochastic process without recourse to differential equations. If we equate the differential coefficients on the left-hand sides of equations (2) to zero and write now p_A, p_B as a shortened form of $p_A(\infty)$, $p_B(\infty)$, then we have the so-called equilibrium equations

$$-\alpha p_A + \beta p_B = 0, \qquad \alpha p_A - \beta p_B = 0. \tag{5}$$

In the special case we are considering here the two equilibrium equations are essentially identical; but this does not happen where there are more than two states. There is a unique solution to (5) which satisfies the probability requirement $p_A + p_B = 1$, namely the equilibrium probability distribution

$$p_A = \frac{\beta}{\alpha + \beta}; \qquad p_B = \frac{\alpha}{\alpha + \beta}.$$

Thus even for the more difficult processes to which we have alluded above, when we are unable to solve the differential equations for compact solutions like (3), we have some hope of obtaining the equilibrium probability distribution, at least, by by-passing the intractable differential equations of the process and writing down the equilibrium equations, analogous to (5), directly. A little practice soon allows one to write down appropriate equilibrium equations quite rapidly.

There are two further important properties of the equilibrium probability distribution. First, suppose that we assign the limiting values $p_A(\infty)$, $p_B(\infty)$ to the initial probabilities $p_A(0)$, $p_B(0)$. Then it is clear from (3) that $p_A(t) = \beta/(\alpha + \beta)$, $p_B(t) = \alpha/(\alpha + \beta)$, for all t, i.e. the instantaneous probability distribution is constant, or *stationary*, independent of time. This can also be seen from equations (2) and (5) and the fact that $p_A(\infty)$, $p_B(\infty)$ satisfy (5).

Because of this property of the equilibrium distribution it is sometimes called the *stationary distribution*. We shall use both terminologies.

The second property is the one that gives the stationary distribution its great practical importance. To explain the property, suppose for simplicity, that the system is initially in state A. The development of the system is represented in Fig. 2.1. Each period spent in a state is a random variable. The random variable has mean α^{-1} for state A, and β^{-1} for state B. In a very long time,

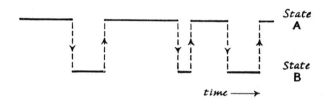

Fig. 2.1. The two-state process. Individual times spent in state A are random variables of mean α^{-1}; those for state B are random variables of mean β^{-1}.

after $2N$ transitions, say, the total time spent in state A will be the sum of a large number N of quantities, each with expectation α^{-1}, and hence will be asymptotically* $N\alpha^{-1}$; similarly the time spent in state B will be asymptotically $N\beta^{-1}$. Therefore the proportion of a very long time spent in state A is asymptotically

$$\frac{N\alpha^{-1}}{N\alpha^{-1}+N\beta^{-1}} = \frac{\beta}{\alpha+\beta} = p_A. \tag{6}$$

Thus, if we observe the process for a long period of time the proportion of time which the system spends in state A is approximately the stationary probability p_A of finding the system in state A. Similarly for state B.

* A rigorous development of this argument uses a law of large numbers from the theory of probability.

D

Note that the argument leading to (6) requires a knowledge of only the mean duration of the stay in each state, whereas the equations (1)–(5) require exponential distributions. For most of the systems that we shall discuss a limited amount of information, independent of distributional assumptions, can be obtained by direct consideration of long-run behaviour, but the full solution, even for the equilibrium distribution, will require assumptions about the frequency distribution of service-time, etc.

In many practical applications the average of some aspect of the system over a fairly long time period is required, and this can be estimated from the equilibrium probability distribution, if it exists. Two general points have to be watched. One is that the initial conditions must not be too extreme; the other is that the time period must be sufficiently long for the observed average to be near its expected value with high probability. Both points can be investigated mathematically in sufficiently simple cases.

Let us now summarize the results of this section. We shall do this in terms of a more general stochastic process than the very special one we have studied because our results are, in fact, true for quite a wide range of processes. We suppose, then, that we have a set of 'states' A_0, A_1, ..., and we write $p_n(t)$ for the probability of finding the system in state A_n at time t. We also suppose that if the system is in state A_n at any time t then there is a probability $\alpha_n \delta t$ (to the first order of differentials) of the system passing to A_{n+1} in the time-interval $(t, t+\delta t)$, independent of the history of the system before time t; this last requirement is crucial. There is likewise a probability $\beta_n \delta t$ of the system passing to state A_{n-1}; of course, we must have $\beta_0 = 0$ because there is no state A_{-1}.

The probabilities $p_n(t)$ satisfy a system of ordinary differential equations, analogous to (2), of which the typical member is

$$\frac{dp_n(t)}{dt} = -(\alpha_n + \beta_n)p_n(t) + \alpha_{n-1}p_{n-1}(t) + \beta_{n+1}p_{n+1}(t). \quad (7)$$

It may happen that as $t \to \infty$ the instantaneous probability distribution $\{p_0(t), p_1(t), \ldots\}$ approaches an equilibrium or stationary distribution $\{p_0, p_1, \ldots\}$, which does not involve the initial probability distribution $\{p_0(0),\ p_1(0),\ \ldots\}$. If so, the stationary distribution can be found by solving the equilibrium equations

$$\left.\begin{aligned}
\alpha_0\, p_0 &= \beta_1\, p_1, \qquad (\alpha_1+\beta_1)\, p_1 = \alpha_0\, p_0 + \beta_2\, p_2, \\
\text{and, in general,} & \\
(\alpha_n+\beta_n)\, p_n &= \alpha_{n-1}\, p_{n-1} + \beta_{n+1}\, p_{n+1} \quad (n \geqslant 1).
\end{aligned}\right\} \tag{8}$$

Equations (8) are obtained by equating to zero the right-hand sides of (7); their solution is, of course, subject to the normalizing requirement for a probability distribution, namely $p_0 + p_1 + \ldots = 1$.

If the initial probability distribution is given by the stationary distribution, i.e. if $p_n(0) = p_n$, for all n, then the instantaneous distribution $p_n(t)$ has the constant, stationary, value p_n for all t. In general the stationary distribution, if it exists, approximates well to the instantaneous probability distribution $p_n(t)$ for large values of t. It also approximates the relative proportions of a long time stretch which the system spends in the various states.

Finally it must be emphasized that, especially if the number of states is infinite, an equilibrium distribution may not exist. In practice such non-existence nearly always means that under the assumptions of the model the state number tends to increase indefinitely as t increases. For example if the mean service-time exceeds the mean interval between successive arrivals in a single-server queue, the number of customers waiting in the system increases indefinitely with time. There are more pathological possibilities, however. The general theory of Markov processes deals with these matters, but here we shall take the naïve view that if the formal equilibrium equations have a unique solution representing a probability distribution, then this is the equilibrium distribution, having the properties set out above. Conversely, if there is no such unique solution, there is no equilibrium distribution. We return briefly to this point in section 2.4.

2.3 Single-server queue with random arrivals and exponential service-times

Suppose that we have a queueing situation in which a single server deals with a stream of customers arriving at random (section 1.3

(ii)). We shall write α, as before, for the average rate of arrivals; thus the interval of time between two successive arrivals is a random variable with an exponential distribution and mean value α^{-1}. Furthermore, the probability that a fresh customer arrives in a time-interval $(t, t + \delta t)$ is, to the first order of differentials, $\alpha \delta t$ regardless of all that happened prior to the time t.

Suppose further that customers' individual service-times are independent values from a fixed exponential distribution with mean σ^{-1}. Thus, by the discussion of section 1.4 (i) (b), if the server is busy serving a customer at time t, then the probability this customer's service is completed in the interval $(t, t + \delta t)$ is $\sigma \delta t$, whatever the amount of time the server has already expended on this particular customer.

It is convenient to introduce at this stage the notion of *traffic intensity*. For a queueing situation such as we are at present considering, this is a dimensionless quantity, defined as

$$\frac{\text{Mean service-time of a single customer}}{\text{Mean interval between arrivals of successive individual customers}}$$

(9)

and is generally denoted by ρ. Thus $\rho = \alpha/\sigma$ in our present notation. The unit of traffic intensity is called the *erlang*, out of deference to A. K. Erlang who was a pioneer investigator in congestion theory.

Let us now apply the ideas of section 2.2 to our queueing situation. If there are n customers in the queue, *including the one being served*, we say that state A_n is occupied. The probability differential for a transition $A_n \rightarrow A_{n+1}$ is evidently $\alpha \delta t$, and for a transition $A_n \rightarrow A_{n-1}$ we have, similarly, $\sigma \delta t$. The equilibrium equations analogous to (8) then become

$$\left.\begin{aligned}
\alpha p_0 &= \sigma p_1 \\
(\alpha + \sigma)\, p_1 &= \alpha p_0 + \sigma p_2 \\
\cdots \quad &\quad \cdots \\
(\alpha + \sigma)\, p_n &= \alpha p_{n-1} + \sigma p_{n+1} \\
\cdots \quad &
\end{aligned}\right\}$$

(10)

with the normalizing condition

$$p_0 + p_1 + \ldots = 1.$$

If we solve (10) systematically we find that $p_n = \rho^n p_0$. On substituting this into the normalizing equation, we have that

$$p_0 \sum_{n=0}^{\infty} \rho^n = 1,$$

from which $p_0 = 1 - \rho$, and hence in general   if $|\rho| < 1$

$$p_n = (1 - \rho)\rho^n. \tag{11}$$

Now (11) represents a probability distribution if and only if $\rho < 1$ For if $\rho > 1$, the terms of (11) are negative, and if $\rho = 1$ they are all zero. Hence we have a stationary distribution if and only if $\rho < 1$. This condition is intuitively sensible. For if $\rho < 1$ then, on the average, the server is able to deal with more than one customer's requirements before the next customer arrives and we should therefore expect the server to cope satisfactorily with his task. Thus we expect the queue rarely to attain an excessive length. On the other hand, if $\rho > 1$, it is clear that the server will be unable to deal with customers as fast as they arrive and that very long queues will develop. A qualitative account of what happens when $\rho = 1$ is more delicate.

We now consider the consequences of (11). Since $p_0 = 1 - \rho$, the traffic intensity ρ represents that proportion of the time in which the server is busy serving; the remainder of the time he is idle, with no customers to serve. Direct consideration of long-run behaviour, as in section 2.2, shows this to be true for any single-server queue in statistical equilibrium, with ρ defined by (9).

The distribution (11) is a standard distribution in probability theory, called the geometric distribution. Its mean is $\rho/(1 - \rho)$ and its variance is $\rho/(1 - \rho)^2$. The probability of finding more than N customers in the queue is ρ^{N+1}. Thus with low values of the traffic intensity long queues are extremely unlikely. Table 2.1 shows the

TABLE 2.1

Dependence of various characteristics of the queue upon the traffic intensity; single server with random arrivals and exponential service-times.

Traffic intensity ρ	Probability server free $1-\rho$	Mean queue size $\rho/(1-\rho)$	Probability of more than four customers in queue ρ^5
0·1	0·9	0·111	0·00001
0·2	0·8	0·250	0·0003
0·3	0·7	0·429	0·002
0·4	0·6	0·667	0·010
0·5	0·5	1·000	0·031
0·6	0·4	1·500	0·078
0·7	0·3	2·333	0·168
0·8	0·2	4·000	0·328
0·9	0·1	9·000	0·590

dependence of various characteristics of the queue upon the traffic intensity. It will be noticed that the reduction of mean queue size, and of the probability of very long queues, necessarily raises the proportion of time which the server spends idle.

This illustrates a general feature of congestion systems, namely that finding the most economical arrangement of the system usually involves compromising between ensuring full use of the service facilities and minimizing delays to customers.

The above results refer to the number of customers in the system. If we assume that the queue-discipline is 'first-come, first served', then it is possible to infer the probability distribution of the time a typical customer spends in the queue. We defer this calculation until section 2.6, however, where we shall see that it can be performed for quite general service-time distributions.

2.4 Arrival rates and service-times dependent on queue size

The results of the preceding section can easily be extended to cover more general circumstances where, for instance, the rate at which customers arrive may depend on the size of the queue. A long queue may deter customers from joining it and enduring the necessarily long wait for the required service. Or a short queue, by drawing interest to the service available, may tend to attract customers. This latter phenomenon can be observed before exhibits in a trade fair, for example. Further, the rate at which the server deals with his customers may depend upon the size of the queue before him. One possibility is that a long queue causes the server to work more quickly, whereas a very short queue tempts him to work at a more relaxed pace.

Thus we shall suppose, for the present discussion, that transitions $A_n \rightarrow A_{n+1}$ are associated with a probability differential $\alpha_n \delta t$ and that transitions $A_n \rightarrow A_{n-1}$ (for $n > 0$) are associated with a probability differential $\sigma_n \delta t$. We continue to make the strong assumption that the probabilities referring to transitions in $(t, t+\delta t)$ do not depend on what happened before t.

The equilibrium equations giving the stationary distribution of queue-size are the general equilibrium equations (8) of section 2.2. These may be solved systematically, and we find

$$p_1 = \frac{\alpha_0}{\sigma_1} p_0, \qquad p_2 = \frac{\alpha_0 \alpha_1}{\sigma_1 \sigma_2} p_0,$$

and, in general

$$p_n = \frac{\alpha_0 \alpha_1 \dots \alpha_{n-1}}{\sigma_1 \sigma_2 \dots \sigma_n} p_0. \tag{12}$$

We must choose p_0 in (12) so that the resulting sequence $\{p_n\}$ sums correctly to unity. This will only be possible if the series

$$S = 1 + \frac{\alpha_0}{\sigma_1} + \frac{\alpha_0 \alpha_1}{\sigma_1 \sigma_2} + \dots + \frac{\alpha_0 \alpha_1 \alpha_2 \dots \alpha_{n-1}}{\sigma_1 \sigma_2 \sigma_3 \dots \sigma_n} + \dots$$

converges to a finite value. What are we to infer if $S = \infty$? When this happens we are faced with a situation similar to the one encountered in section 2.3 when the traffic intensity was not less than unity. If $S = \infty$ the customers, on the average, are arriving more quickly than the server can deal with them and arbitrarily large queues result. There is no stationary probability distribution of queue size.

In what follows we assume that $S < \infty$. The stationary probability distribution of queue size is then given by *

$$p_n = S^{-1} \frac{\alpha_0 \alpha_1 \ldots \alpha_{n-1}}{\sigma_1 \sigma_2 \ldots \sigma_n} \ (n = 1, 2, \ldots), p_0 = S^{-1}. \quad (13)$$

We now consider, briefly, a number of easy and interesting applications of this formula.

(i) *Queue with discouragement*

Suppose that the sight of a long queue discourages fresh customers from joining it. A simple model would be $\alpha_n = \alpha/(n+1)$, $\sigma_n = \sigma$, independently of n. Then

$$S = 1 + \frac{\alpha}{\sigma} + \frac{\alpha^2}{2! \, \sigma^2} + \ldots = e^{\alpha/\sigma},$$

and we find that

$$p_n = e^{-\alpha/\sigma} \frac{(\alpha/\sigma)^n}{n!} \ (n = 0, 1, 2, \ldots).$$

Thus there is a Poisson distribution of queue size, with mean α/σ. The probability that the server is free is $e^{-\alpha/\sigma}$.

(ii) *Queue with ample servers*

Suppose we have customers arriving at random with the usual average rate α, and suppose, also as usual, that the service-times

* We are here suggesting that there is a stationary probability distribution of queue size if and only if $S < \infty$. As a simple working rule in a book of the present character this should be useful enough. However, the reader is warned that if he applies it to weird set-ups then he may be misled; he should refer to the literature on Markov chains, especially a paper by Ledermann and Reuter (*Proc. Camb. Phil. Soc.* **49** (1953), 247), for guidance. Very roughly, the 'working rule' is satisfactory when α_n/n and β_n/n are bounded functions of $n > 0$.

have a negative exponential distribution with mean σ^{-1}. Suppose however that whenever a customer arrives there is a server made available to deal with him without delay. Thus, when we say the queue-size is n we mean here that there are n customers in the shop, each being served by his own server. Then $\alpha_n = \alpha$, for all n, and $\sigma_n = n\sigma$. This leads to $S = e^{\alpha/\sigma}$, and again to a Poisson distribution of queue size. The reader familiar with stochastic processes considered in other fields will see that this example is equivalent to the immigration-death process.

(iii) *Queue with m servers*

To make the previous example smack more of reality, consider the modification which arises by imposing an upper limit m on the number of servers that can be made available. Thus, if $n \leqslant m$, a 'queue' of n customers would really consist of n customers simultaneously being served. But if $n > m$ then a 'queue' of n customers would consist of m customers simultaneously being served together with a genuine queue of $n - m$ waiting customers. For this set-up we have $\alpha_n = \alpha$, for all n, and $\sigma_1 = \sigma$, $\sigma_2 = 2\sigma$, ..., $\sigma_{m-1} = (m-1)\sigma$, $\sigma_n = m\sigma$, for all $n \geqslant m$. Thus, if $\alpha < m\sigma$,

$$S = 1 + \frac{\alpha}{\sigma} + \frac{\alpha^2}{2!\,\sigma^2} + \ldots + \frac{\alpha^{m-1}}{(m-1)!\,\sigma^{m-1}} + \frac{m^{m-1}}{(m-1)!} \sum_{n=m}^{\infty} \left(\frac{\alpha}{m\sigma}\right)^n.$$

If we write $\rho = \alpha(m\sigma)^{-1}$ we find that

$$S = 1 + (m\rho) + \frac{(m\rho)^2}{2!} + \ldots + \frac{(m\rho)^{m-1}}{(m-1)!} + \frac{(m\rho)^m}{m!(1-\rho)}.$$

It then transpires that

$$p_n = S^{-1} \frac{(m\rho)^n}{n!} \quad (n < m)$$

and

$$p_n = S^{-1} \frac{m^m}{m!} \rho^n \quad (n \geqslant m).$$

There are no compact expressions for means, variances, etc., of the resulting stationary distribution of queue-size; but for a given value of m the necessary computations are not particularly troublesome.

For example, with $m = 2$, $S = (1 + \rho)/(1 - \rho)$, and we find that

$$p_0 = (1 - \rho)/(1 + \rho),$$

$$p_n = 2\rho^n(1 - \rho)/(1 + \rho) \quad (n \geq 1).$$

We also find that the mean queue size is $2\rho/(1 - \rho^2)$. It is interesting to compare these results with the results of section 2.3 for $m = 1$.

(iv) *Queue with limited waiting room*

Suppose we have the same situation as envisaged in section 2.3, i.e. single server, random arrivals, and exponential service-times, but suppose that there is no room for more than R customers to queue (including the one being served). Here we have

$$\alpha_n = \alpha \quad (n < R),$$
$$= 0 \quad (n \geq R),$$
$$\sigma_n = \sigma \quad (n = 1, 2, \ldots).$$

This set of equations implies that customers who arrive and find no waiting room leave without service and do not return later. Then S is the finite sum

$$1 + \frac{\alpha}{\sigma} + \left(\frac{\alpha}{\sigma}\right)^2 + \ldots + \left(\frac{\alpha}{\sigma}\right)^R,$$

because of the vanishing of α_R. We can then show that for $n \leq R$

$$p_n = \frac{\left(\dfrac{\alpha}{\sigma}\right)^n \left\{1 - \dfrac{\alpha}{\sigma}\right\}}{\left\{1 - \left(\dfrac{\alpha}{\sigma}\right)^{R+1}\right\}}.$$

(v) *Calls on telephone exchange with k lines*

Suppose that we have k servers and waiting room for only k customers. This is exactly the situation which arises when we consider a telephone exchange with just k trunk lines and no facility for holding subscribers who require a line but cannot be supplied with one. It is assumed that such calls are lost.

Here we evidently have

$$\alpha_n = \alpha \quad (n < k),$$
$$= 0 \quad (n \geq k),$$
$$\sigma_n = n\sigma \quad (n \leq k).$$

(There is no point in defining σ_n for $n > k$.) As in (iv) we find that S is a finite sum:

$$S = 1 + \frac{\alpha}{\sigma} + \frac{1}{2!}\left(\frac{\alpha}{\sigma}\right)^2 + \ldots + \frac{1}{k!}\left(\frac{\alpha}{\sigma}\right)^k,$$

and that for $n \leq k$,

$$p_n = \frac{\frac{1}{n!}\left(\frac{\alpha}{\sigma}\right)^n}{1 + \left(\frac{\alpha}{\sigma}\right) + \frac{1}{2!}\left(\frac{\alpha}{\sigma}\right)^2 + \ldots + \frac{1}{k!}\left(\frac{\alpha}{\sigma}\right)^k}.$$

The quantity p_k gives the proportion of time for which the system is fully occupied and in the telephone application, with randomly arriving calls, this is the proportion of lost calls. The formula for p_k is often called Erlang's loss formula.

(vi) *Machine-minding*

Suppose that we have k machines under the care of a single operator. Suppose that if a given machine is running at time t the probability that it breaks down in $(t, t+\delta t)$ is $\alpha\delta t$, to the first order. Suppose that a repair-time for a machine can be regarded as a random sample from an exponential distribution with mean σ^{-1}. Suppose that no time is lost in the single operator's moving from the location of one machine to that of another, and that he keeps working whenever there is a machine stopped. This set-up can be thought of as a queue in which stopped machines represent customers awaiting service. Furthermore, the longer the 'queue' the slower the arrival of fresh 'customers'. In fact, we have

$$\alpha_n = (k-n)\alpha \quad (n \leq k)$$
$$= 0 \quad (n > k)$$
$$\sigma_n = \sigma \quad (n = 1, 2, \ldots).$$

Thus

$$S = 1 + k\left(\frac{\alpha}{\sigma}\right) + k(k-1)\left(\frac{\alpha}{\sigma}\right)^2 + \ldots + k!\left(\frac{\alpha}{\sigma}\right)^k;$$

and we have $p_0 = S^{-1}$ and, for $n = 1, 2, \ldots, k$,

$$p_n = \frac{k(k-1)\ldots(k-n+1)\left(\dfrac{\alpha}{\sigma}\right)^n}{1 + k\left(\dfrac{\alpha}{\sigma}\right) + k(k-1)\left(\dfrac{\alpha}{\sigma}\right)^2 + \ldots + k!\left(\dfrac{\alpha}{\sigma}\right)^k}.$$

A fuller discussion of such machine-minding problems will be given in Chapter IV.

Note that although the α_n, σ_n for the present problem and for (v) are not the same, the resulting p_n are closely related. This fact can be given a physical interpretation.

2.5 Further remarks on equilibrium conditions

In section 2.3 we obtained $\rho < 1$ as the condition for the existence of an equilibrium distribution. In the present section we make some general qualitative comments on the interpretation of such conditions and then discuss briefly a more theoretical matter, the definition of traffic intensity for the more general process of section 2.4, and its relation to the existence of an equilibrium distribution.

The practical interpretation of an equilibrium condition has to be made with caution. In a system in which both arrival pattern and service-time show little random variation it is possible for a system that is theoretically not in equilibrium to remain in apparent equilibrium over quite a long period. In a system with a good deal of random variation, it is possible for the observed behaviour over long time periods to be different from that given by the equilibrium distribution. Both points are covered by the remark that we are always interested in the behaviour of the system over a *finite* time interval, usually, however, long compared with individual service and arrival times. In general, equilibrium theory gives an adequate approximation, but, especially for traffic intensities near unity, we have to be careful. Furthermore, near the critical point, minor effects, such as slight dependencies of arrival rate on queue size, may be important in determining whether or not an equilibrium distribution exists.

For most of the congestion processes considered in this monograph, definition (9) of section 2.3 gives directly a meaningful value of the traffic intensity. The question arises, however, as to how we should define the traffic intensity for the more general class of congestion processes considered in section 2.4. A suitable definition seems to be that the traffic intensity ρ is equal to the reciprocal of the radius of convergence of the power series

$$\sum_{n=1}^{\infty} \frac{\alpha_0 \alpha_1 \ldots \alpha_{n-1}}{\sigma_1 \sigma_2 \ldots \sigma_n} z^n. \tag{14}$$

This definition agrees with that of section 2.3, as the reader may easily verify.

If the radius of convergence of (14), R say, is strictly less than unity then the series S, which may be obtained by putting $z = 1$ in (14), diverges. Thus if $\rho = R^{-1} > 1$ we have no stationary probability distribution. Similarly, if $R > 1$, S will be finite; hence $\rho < 1$ implies the existence of a stationary probability distribution. Thus (14) gives ρ the right properties for agreement with the more familiar queueing situations.

In section 2.3 we found that in the single-server queue with random arrivals and exponential service-time no equilibrium distribution exists when $\rho = 1$. This result applies widely to single-server queues with no dependence between arrival rate, service-time, and queue size. Nevertheless, this need not be so for the situation of section 2.4, as is shown by the following example.

Suppose that $\alpha_n = (n+1)/(n+2)$, $\sigma_n = (n+1)/n$, for all n. This would represent a queue in which an increase in queue-length produces a very slight increase in the arrival rate, which rate is very nearly unity for fairly long queues. Further the server tends to work slightly slower when the queue gets longer, for example his time and attention may be partially involved with regulating the queue in some way; but for fairly long queues the mean service-time is very nearly unity. Our definition assigns the value unity to the traffic intensity for this set-up. However, the series $\sum n^{-2}$ is convergent, so that there is a stationary distribution of queue-size even though $\rho = 1$.

2.6 Single-server queue with random arrivals and general distribution of service-times

Section 2.3 dealt with a single-server queue with random arrivals and an exponential distribution of service-time. For many applications the assumption of random arrivals is reasonable; however, the assumption of exponential distributions of service-time is often unsatisfactory. It is therefore worthwhile to investigate the single-server queue with random arrivals and with the service-times of different customers independently distributed, with a distribution function $B(x)$, which is not necessarily of a particularly simple mathematical form. As a special case the service-time may be constant. It is assumed that service-times and arrival-times do not depend on the number of customers queueing. If the server is busy serving at time t we can no longer suppose, as we did in section 2.3, that the probability he finishes dealing with his present customer in $(t, t+\delta t)$ is given by the simple expression $\delta t/b_1$, where b_1 denotes the mean service-time. For if we made such an assumption, it would follow, as in section 1.4 (i) (b), that service-times have an exponential distribution and we would not have achieved the generality we now desire. Thus we can no longer use the simple 'δt' technique and therefore we cannot write down the simple equilibrium equations either.

There are several possible methods for getting over this difficulty; the one we follow here depends on examining the system at suitably selected isolated time points. Suppose then that at the initial moment $t = 0$ there are q_0 customers in the queue. Let the customers be in line, numbered $1, 2, ..., q_0$ from the head of the queue, and suppose that service is just about to commence for customer number 1. As further customers arrive and join the queue, number them $q_0 + 1, q_0 + 2, ...$ and so on, in an obvious manner. We shall then write q_n for the number of people in the queue at the moment the service of customer number n terminates, and ξ_n for the number of customers that arrive during the service-time of customer number n. The quantities q_n, ξ_n are random variables and evidently $\xi_1, \xi_2, ...$ are independent and identically distributed. Furthermore ξ_n is independent of $q_0, q_1, ..., q_{n-1}$. Our discussion of the single-server queue with random arrivals and general independent service-times will be based on the sequence of random variables $\{q_n\}$.

The development of the sequence $\{q_n\}$ is summed up in the equation

$$q_{n+1} = \begin{cases} q_n - 1 + \xi_{n+1} & (q_n > 0) \\ \xi_{n+1} & (q_n = 0). \end{cases}$$

The first equation, for $q_n > 0$, says that the queue left behind by customer $n + 1$ consists of the queue left behind by customer n, diminished by one customer, customer number $n + 1$, and augmented by the customers who arrive during the service-time of customer $n + 1$. The equation for $q_n = 0$ is based on the reasoning that customer $n + 1$ must arrive to find the server free, so that the queue he leaves behind him consists only of those customers who arrive during his service-time.

If we define Heaviside's Unit Function $U(x)$ by

$$U(x) = \begin{cases} 1 & (x > 0), \\ 0 & (x \leqslant 0), \end{cases}$$

both equations for q_{n+1} can be combined into the one equation

$$q_{n+1} = q_n - U(q_n) + \xi_{n+1}. \tag{15}$$

$$U(q_n)\left[q_n - 1\right] + \xi_{n+1}$$

To use (15) we must know the distribution of the ξ_n. If the service-time of customer n is x, then ξ_n has a Poisson distribution of mean αx, that is

$$\text{prob}(\xi_n = k \,|\, x) = \frac{e^{-\alpha x}(\alpha x)^k}{k!}.$$

But the service-time has distribution function $B(x)$, so that

$$\eta_k = \text{prob}(\xi_n = k) = \int_0^\infty \frac{e^{-\alpha x}(\alpha x)^k}{k!} \, dB(x).$$

The probability generating function of ξ_n is

$$\left. \begin{aligned} \Xi(\zeta) &= \sum_{k=0}^\infty \zeta^k \text{prob}(\xi_n = k) \\ &= \int_0^\infty e^{-\alpha x(1-\zeta)} \, dB(x) \\ &= B^\star(\alpha - \alpha\zeta), \end{aligned} \right\} \tag{16}$$

where

$$B^\star(s) = \int_0^\infty e^{-st} \, dB(t)$$

is the Laplace–Stieltjes transform of the distribution of service-time. The moments of service-time are given by differentiating $B^\star(s)$ at $s = 0$, and the factorial moments of ξ_n by differentiating $\Xi(\zeta)$ at $\zeta = 1$. In fact

$$\begin{aligned} E[\xi_n(\xi_n - 1)\dots(\xi_n - r + 1)] &= \Xi^{(r)}(1) \\ &= (-\alpha)^r B^{\star(r)}(0) \\ &= \alpha^r b_r, \end{aligned}$$

where b_r is the rth moment about the origin of service-time. In particular

$$E\xi_n = \alpha b_1 = \rho \tag{17}$$

$$\text{var}(\xi_n) = \alpha b_1 + \alpha^2 \sigma_b^2 = \rho + \rho^2 C_b^2, \tag{18}$$

where σ_b^2, C_b^2 are the variance and fractional coefficient of variation squared of service-time.

Now that the distribution of the random variables ξ_n is known, one can determine from (15) the distribution of q_1 in terms of q_0, the distribution of q_2 in terms of q_1, and so on, by putting $n = 0, 1, 2, \ldots$, etc. This gives us in principle the properties of the system at the particular sequence of time instants under consideration. While this does not answer all questions that might be asked about the system, it is enough for many purposes.

As we would expect from our treatment of the continuous-time processes in earlier sections, if the traffic intensity, $\rho = \alpha b_1 < 1$, a stationary probability distribution $\{\pi_k\}$ exists, that is

$$\lim_{n \to \infty} \text{prob}(q_n = k) = \pi_k.$$

The probability π_k also has the interpretation that it approximates the proportion of a large number of successive customers who leave behind them a queue of size k. Furthermore, if the random variable q_0 is assigned the stationary probability distribution $\{\pi_k\}$, then

$$\text{prob}(q_n = k) = \pi_k$$

for all n. We then say that the queueing process is stationary.

To determine the stationary distribution we use the key equation (15) together with the properties of the random variables ξ_n. In the stationary state the probability properties of q_n and of q_{n+1} are identical. Hence in particular $Eq_n = Eq_{n+1}$ and on taking expectations in (15), we have that

$$EU(q_n) = \rho. \tag{19}$$

But $EU(q_n) = \text{prob}(q_n \neq 0)$, so that when the queueing process is stationary the probability that a customer leaves behind him a free server (faced with no customers) is $1 - \rho$. Since for each customer who leaves behind a queue of zero length there must be just one customer who arrives to find no queue ahead of him, we can infer that the probability that a customer does not have to wait in the queue is also $1 - \rho$. A related result, that can be proved

E

by direct consideration of long-run averages, is that the system is empty for a proportion $1 - \rho$ of the time.

Equation (15) will also yield information concerning the mean queue-size, Eq_n, fairly easily. If we square both sides of (15), noting that $qU(q) = q$, and take expectations, we obtain

$$Eq_{n+1}^2 = Eq_n^2 + EU(q_n) + E\xi_{n+1}^2 + 2Eq_n\xi_{n+1}$$
$$- 2E\xi_{n+1}U(q_n) - 2Eq_n.$$

Since Eq_n^2 does not involve n, in the stationary case. and since ξ_{n+1} and q_n are independent, we deduce that

$$Eq_n = \frac{\rho + E\xi_{n+1}^2 - 2\rho E\xi_{n+1}}{2[1 - E\xi_{n+1}]}$$

$$= \frac{\rho + \rho + \alpha^2\sigma_b^2 + \rho^2 - 2\rho^2}{2(1-\rho)}$$

$$= \rho + \frac{\rho^2(1 + C_b^2)}{2(1-\rho)}. \tag{20}$$

This formula is a most useful one when discussing the effects of proposed changes in a service-mechanism. Notice that even if the mean service-time cannot be decreased one can reduce the average queue-size by cutting down on the variability of service-times, i.e. by reducing σ_b^2. If service is exponential, $C_b^2 = 1$ and (20) reduces to

$$Eq_n = \rho + \frac{2\rho^2}{2(1-\rho)} = \frac{\rho}{1-\rho} \tag{21}$$

On the other hand if service-time is constant, $C_b = 0$, and

$$Eq_n = \frac{\rho(1 - \tfrac{1}{2}\rho)}{1-\rho}$$

and the ratio of this to (21) is $1 - \tfrac{1}{2}\rho$. Thus if ρ is small there is little reduction in mean queue size in changing from $C_b = 1$ to $C_b = 0$ and a reduction approaching $\tfrac{1}{2}$ as $\rho \to 1$.

An interesting aspect of (21) is the following. We found in

section 2.3 that for random arrivals and exponential service-times the single-server queue in its stationary state had a mean queue-size given by $\rho/(1-\rho)$. This result referred to an arbitrary moment in time. In this section we have studied the same process but focused our attention on those instants when a customer leaves the queue, his service completed. We find that the mean queue-size left behind by such a customer is, when the queueing process is stationary, also $\rho/(1-\rho)$. The equality of these two mean queue-sizes is hard to reconcile intuitively, but is a fact nevertheless. It is a property of queues with random arrivals only, and is not found when more general arrival patterns are studied.

To find the mean waiting-time, note that if the waiting-time of the nth customer is w_n, then with the queue-discipline 'first-come, first-served', q_n is the number of customers which arrive during a time interval of length w_n. Thus $Eq_n = \alpha Ew_n$, and we have at once that the mean waiting-time is

$$Ew_n = \frac{1}{\alpha}\left\{\rho + \frac{\rho^2(1+C_b^2)}{2(1-\rho)}\right\},$$

or

$$\frac{Ew_n}{b_1} = 1 + \frac{\rho(1+C_b^2)}{2(1-\rho)}. \tag{22}$$

This is often called Pollaczek's formula. The dependence on the coefficient of variation of service-time, C_b, is the same as in (20).

So far we have obtained only the average values of variables of interest. To obtain the stationary distribution of queue-size, write $Q(\zeta) = E\zeta^{q_n}$ for its probability generating function. Then from (15)

$$\begin{aligned} Q(\zeta) &= E\zeta^{q_n - U(q_n) + \xi_{n+1}}, \\ &= \Xi(\zeta)\,E\zeta^{q_n - U(q_n)}, \end{aligned} \tag{23}$$

since ξ_{n+1} is independent of q_n.

However, we have that

$$E\zeta^{q_n - U(q_n)} = \pi_0 + \sum_{n=1}^{\infty} \pi_n \zeta^{n-1} = (1-\rho) + \frac{Q(\zeta) - (1-\rho)}{\zeta}. \tag{24}$$

If we substitute (24) into (23) and perform some elementary manipulations, we get that

$$Q(\zeta) = \frac{(1-\rho)(1-\zeta)\,\varXi(\zeta)}{\varXi(\zeta)-\zeta}. \tag{25}$$

When we know the service-time distribution and recall that (16) shows that $\varXi(\zeta) = B^\star(\alpha - \alpha\zeta)$, then we can, in principle, expand the right-hand side of (25) and obtain the stationary probability distribution $\{\pi_n\}$. Of course, it will frequently happen that such an expansion presents too formidable a task, but even in such cases the moments of the distribution $\{\pi_n\}$ are fairly easy to obtain.

It is of interest to see the reduction of (25) when service-times are exponential, with mean value σ^{-1}. We then have $B^\star(s) = \sigma/(\sigma+s)$, so that $\varXi(\zeta) = [1+\rho(1-\zeta)]^{-1}$. Substitution in (25) then gives $Q(\zeta) = (1-\rho)/(1-\rho\zeta)$, whence $\pi_n = \rho^n(1-\rho)$. Thus the queue-size at the moments when service is completed has the same stationary geometric distribution as the one which we found applied to the queue-size at an arbitrary moment. This result generalizes the discovery we made earlier concerning the mean queue-size, but is equally difficult to understand intuitively. Further generalizations are possible.

To close this section we return to a consideration of the waiting-time of a customer. We have seen that we can calculate the mean waiting-time from the mean queue-size. In a similar way we can derive the distribution of waiting-time from the distribution of queue-size.

Let $W(w)$ be the stationary distribution function of waiting-time. As we have already remarked, q_n is the number of customers to arrive during a time interval w_n. Thus the relation of $Q(\zeta)$ to $W^\star(s)$ is exactly analogous to the relation we have already discovered between $\varXi(\zeta)$ and $B^\star(s)$. We can therefore infer from (16) that

$$Q(\zeta) = W^\star(\alpha - \alpha\zeta). \tag{26}$$

If we substitute $s = \alpha(1 - \zeta)$ in (26), and make use of (25) and (16), then we find that

$$W^\star(s) = \frac{(1-\rho)\,sB^\star(s)}{s - \alpha + \alpha B^\star(s)} \qquad (27)$$

is the Laplace–Stieltjes transform of the distribution of waiting time.

We shall write v_n for the queueing-time of customer n and $V(v)$ for its stationary distribution function. Then $w_n = v_n + t_n$, where t_n is the service-time of customer n. Evidently v_n and t_n are independent. Thus $W^\star(s) = V^\star(s)\,B^\star(s)$, and we can deduce from (27) that

$$V^\star(s) = \frac{(1-\rho)\,s}{s - \alpha + \alpha B^\star(s)}. \qquad (28)$$

Whether or not these formulae can be inverted explicitly to yield the actual distribution-functions W and V from their transforms $W^\star$ and $V^\star$ very much depends on the special form assumed by $B^\star$. As is to be expected, when service-times have a negative-exponential distribution the inversion presents little difficulty. For then $B^\star(s) = \sigma/(\sigma + s)$ and (27) yields

$$W^\star(s) = \frac{\sigma - \alpha}{\sigma - \alpha + s}. \qquad (29)$$

The distribution corresponding to the Laplace–Stieltjes transform (29) is the exponential distribution with a mean of $1/(\sigma - \alpha)$, so that we have proved that the probability density function of waiting-time is

$$(\sigma - \alpha)\,e^{-(\sigma - \alpha)t}. \qquad (30)$$

A direct proof of (29), and hence of (30), from the results of section 2.3 is instructive. Since arrivals are random, independent of the number of customers in the queue, the probability that a newly arrived customer C, has in front of him n customers is $(1-\rho)\rho^n$. Given that n is zero, C's waiting-time is equal to his service-time and has Laplace transform $\sigma/(\sigma+s)$, corresponding to an exponential distribution of mean $1/\sigma$. Given that n is one, the waiting-time is the sum of the service-time of C, and the remaining service-time of the customer currently at the service-point. These are independent exponential variables and the conditional Laplace transform of the distribution of C's waiting-time is $\sigma^2/(\sigma+s)^2$. In general, with probability $(1-\rho)\rho^n$ the conditional Laplace transform is $\sigma^{n+1}/(\sigma+s)^{n+1}$. Thus the unconditional Laplace transform is

$$\sum_{n=0}^{\infty} (1-\rho)\frac{\rho^n\sigma^{n+1}}{(\sigma+s)^{n+1}}. \tag{31}$$

On using the fact that $\rho = \alpha/\sigma$, (31) reduces immediately to (29).

It can be shown similarly that the queueing-time has a distribution which is a mixture of a discrete probability $1-\rho$ at $t=0$ and a density $\rho(\sigma-\alpha)e^{-(\sigma-\alpha)t}$ $(t \geqslant 0)$.

2.7 The busy-period distribution

In the preceding sections we have obtained for the single-server queue with random arrivals quite general equilibrium expressions for the first two of the three queue properties mentioned in section 1.6, namely the waiting-time and the queue size. There remains the distribution of busy periods. The full calculation of this raises very interesting questions in probability theory, which are dealt with briefly as a specialized topic in section 5.6. It is, however, possible to obtain some simple results by entirely elementary arguments.

The server's time is divided into alternating busy and idle periods. With random arrivals there is a constant chance $\alpha\delta t$ of an idle period being ended by the arrival of a customer. Therefore the distribution of the length of idle periods is exponential with mean α^{-1}. Now, whatever the distribution of service-time, the equilibrium probability that the server is idle is $1-\rho$. In a very long time period T, the server is therefore idle for a time $T(1-\rho)$, and this corresponds to $T\alpha(1-\rho)$ distinct idle periods. There are therefore $T\alpha(1-\rho)$ distinct busy periods occupying a total time $T\rho$. This shows that the mean length of a busy period is

$$T\rho/[T\alpha(1-\rho)] = b_1/(1-\rho),$$

where b_1 is the mean service-time. Similarly it can be shown that the mean number of customers served per busy period is $(1-\rho)^{-1}$.

CHAPTER III

More about Simple Queues

3.1 Non-equilibrium theory

In the last chapter we are concerned solely with limiting, or equilibrium, probability distributions. These give the behaviour of the system at a remote future time, and also provide information about the properties of the system averaged over long time periods. Quite often, however, we want more explicit information about the behaviour of the queue over fairly short times. This leads us to the non-equilibrium theory of queues, which is usually much more difficult mathematically than the equilibrium theory.

The following are some examples where non-equilibrium theory is needed:

(a) If the rate of arrivals is suddenly increased for a period of time R, how will the waiting-times increase, and how long must elapse after the end of this period of heavy demand before the situation can be described as 'normal' again?

(b) If there are several servers, and waiting-times are at present low, and it is desired to rest some servers (or to overhaul them if the servers happen to be machines), how many servers may be rested, and for how long, before waiting-times reach prohibitive lengths?

(c) If a queueing system does not run continuously, as supposed in the equilibrium theory, but starts up every morning, say, in a pre-arranged condition (e.g. with the server free and no customers queueing), what can one say about average waiting-times, etc.? If service-times are so long that a server only deals with a few customers each day, an uncritical application of equilibrium theory may be quite misleading.

We quote only these three general problems from a host of possibilities, many of which will doubtless occur to the reader. Problem (*a*) arises when we want to determine the effect of a 'rush-hour' upon the system. Problem (*b*) arises whenever the service mechanism needs occasional checks, rests or overhauls, and a reserve of servers is not available. Problem (*c*) needs little further comment; it is relevant, for example, to the study of appointment systems in hospital outpatient departments, where a given 'clinic' runs for only an hour or two.

All three problems quoted above, and especially (*a*) and (*b*), need a careful study of the *transient* behaviour of the system; equilibrium theory is of little use in their resolution. In this section we shall have to content ourselves to some extent with the solution of rather simpler problems than those just outlined, because a full study would need more space and more elaborate mathematics than we wish to use.

Consider a simple queue with a single server, with arrivals at random with rate α, and with service-times exponentially distributed with mean σ^{-1}. It is convenient first to change the time-scale, taking the mean service-time as the time unit; the arrival rate is then equal to $\alpha/\sigma = \rho$, the traffic intensity. For the remainder of this section we work in the new time-scale.

Let $p_n(t)$ denote the probability that there are n customers in the queue at time t, including the one, if any, being served. By (2.7)* we have the following differential equations:

$$\dot{p}_0 = -\rho p_0 + p_1, \tag{1}$$

$$\dot{p}_n = \rho p_{n-1} - (1+\rho)p_n + p_{n+1} \quad (n = 1, 2, \ldots). \tag{2}$$

Equation (1) is different from the general form (2) because of the impossibility of there being a negative number of customers in the queue. It is precisely this difference that makes the system (1) and (2) awkward to solve. To get round the difficulty we use the following trick. Let equation (2) hold for all integral values of n, positive,

* I.e. equation (7) of Chapter II.

zero and negative. Then, if we can find a solution of (2) satisfying the additional requirement that

$$p_0(t) = \rho p_{-1}(t),$$

the system (1) and (2) will be satisfied.

To solve the extended system (2) we use a generating function. Multiply (2) by ζ^n, where ζ is a dummy variable, and sum over n from $-\infty$ to ∞. Then the generating function

$$P(\zeta, t) = \sum_{n=-\infty}^{\infty} \zeta^n p_n(t)$$

satisfies the differential equation

$$\frac{\partial P}{\partial t} = \left[\rho\zeta - (1+\rho) + \frac{1}{\zeta}\right] P,$$

the most general solution to which is

$$P(\zeta, t) = \Phi(\zeta) \exp\left[-(1+\rho)t + (\rho\zeta + 1/\zeta)t\right], \qquad (3)$$

where $\Phi(\zeta)$ is an arbitrary function of ζ.

At this stage we introduce the Bessel functions of imaginary argument, $I_n(y)$, given by the well-known expansion

$$\exp\left[\tfrac{1}{2}y(z+1/z)\right] = \sum_{n=-\infty}^{\infty} I_n(y) z^n. \qquad (4)$$

An important property of these functions, which we shall use shortly, is that $I_n(y) = I_{-n}(y)$; this result is easy to verify by replacing z by $1/z$ in both sides of (4).

If we substitute $y = 2t\sqrt{\rho}$ and $z = \zeta\sqrt{\rho}$ in (4), we have that

$$\exp\left[t(\rho\zeta + 1/\zeta)\right] = \sum_{n=-\infty}^{\infty} I_n(2t\sqrt{\rho}) \zeta^n \rho^{\frac{1}{2}n},$$

so that from (3),

$$P(\zeta, t) = \Phi(\zeta) e^{-(1+\rho)t} \sum_{n=-\infty}^{\infty} I_n(2t\sqrt{\rho}) \zeta^n \rho^{\frac{1}{2}n}.$$

If now we replace $\Phi(\zeta)$ in this last equation by an expansion

$$\sum_{n=-\infty}^{\infty} \psi_{-n} \zeta^n,$$

whose coefficients are arbitrary, and pick out the coefficients of ζ^n from both sides, we find that the most general solution to the extended system (2) is

$$p_n(t) = e^{-(1+\rho)t} \sum_{r=-\infty}^{\infty} \psi_r \rho^{\frac{1}{2}(n+r)} I_{n+r}(2t\sqrt{\rho}). \tag{5}$$

We now choose the coefficients ψ_n so that (5) is the particular solution we want. If initially there are N customers in the queue, we require that

$$p_n(0) = \delta_{Nn} \quad (n = 0, 1, 2, \ldots),$$

where δ_{ij} is the usual Kronecker delta symbol. Note that we are not concerned with $p_n(0)$ for $n < 0$.

On putting $y = 0$ in (4), we have that $I_n(0) = \delta_{n0}$, so that our condition on $p_n(0)$ becomes, from (5),

$$\psi_{-n} = \delta_{Nn} \quad (n = 0, 1, 2, \ldots).$$

Thus the general solution (5) is now specialized to

$$p_n(t) = e^{-(1+\rho)t}\{\rho^{\frac{1}{2}(n-N)} I_{n-N}(2t\sqrt{\rho})$$

$$+ \sum_{r=1}^{\infty} \psi_r \rho^{\frac{1}{2}(n+r)} I_{n+r}(2t\sqrt{\rho})\}. \tag{6}$$

To determine the remaining ψ's we use the condition that $p_0(t) = \rho p_{-1}(t)$. Put $\phi_r = \psi_r \rho^{\frac{1}{2}r}$ and write I_n for $I_n(2t\sqrt{\rho})$. We require that

$$\rho^{-\frac{1}{2}N} I_N + \sum_{r=1}^{\infty} \phi_r I_r = \rho^{-\frac{1}{2}N+\frac{1}{2}} I_{N+1} + \rho^{\frac{1}{2}} \sum_{r=1}^{\infty} \phi_r I_{r-1}. \tag{7}$$

If we compare coefficients of I_n in (7), we have that

$$\phi_r = 0 \quad (r = 1, 2, \ldots, N),$$
$$\phi_{N+1} = \rho^{-\frac{1}{2}N - \frac{1}{2}},$$
$$\phi_{N+k+1} = \rho^{-\frac{1}{2}N - \frac{1}{2}k - \frac{1}{2}} (1 - \rho) \quad (k = 1, 2, \ldots).$$

This determines the solution for $p_n(t)$, namely

$$p_n(t) = \rho^{\frac{1}{2}(n-N)} e^{-(1+\rho)t} \{ I_{N-n} + \rho^{-\frac{1}{2}} I_{n+N+1}$$
$$+ (1-\rho) \sum_{k=1}^{\infty} \rho^{-\frac{1}{2} - \frac{1}{2}k} I_{n+N+1+k} \}. \tag{8}$$

This equation holds both for $\rho \geqslant 1$ and for $0 \leqslant \rho < 1$, so that (8) describes both the indefinite growth of a system with $\rho \geqslant 1$, and also the approach to the equilibrium distribution of a system with $0 \leqslant \rho < 1$. The solution (8) is, however, far from convenient; when we consider that it originates from one of the very simplest queueing systems, the difficulty of obtaining general solutions in more complicated cases will be apparent. Other properties of the system are in principle determined by (8); thus with the queue-discipline first-come, first-served, the distribution of waiting-time for a customer arriving at time t is found by the argument of (2.27).

As an example of how (8) might be applied to a practical non-equilibrium problem, consider problem (*a*) mentioned at the beginning of this chapter. If we suppose our model (single server, random arrivals, etc.) to be appropriate, then quite a detailed study can be made of the effect of a rush-hour (at the expense of some heavy computation!). For we can assume a certain value N for the queue-size at the start of the rush-hour, assign ρ its appropriate high value (presumably greater than 1) and, by means of (8), calculate the exact probability distribution of queue-size for an increasing sequence of time-points. Similarly, one can study the effect of terminating a rush-hour by assigning N a very high value, typical of the values found to exist near the end of the rush period, and letting ρ have its 'usual' modest value.

One can also apply (8) to problem (*c*) if the simple model with exponential service-time and random arrivals is appropriate. We omit further details.

Evidently somewhat less precise arguments must, in general, be adopted, if we are to obtain useful formulae applicable to fairly general non-equilibrium queueing problems. We illustrate one type of argument that can be used by considering again the rush-hour problem. Suppose that we have a single-server queue; that service-times are independently distributed, with mean value b and coefficient of variation C. Suppose that arrival intervals are independently distributed with mean a, that a customer arrived at the initial instant, and that the variance of the number of customers to arrive by time t is $V(t)$. Thus for regular arrivals $V(t) = 0$, whereas for random arrivals $V(t) = t/a$. The traffic intensity ρ is equal to b/a.

Suppose now that there are N_0 customers in the queue initially and that $\rho > 1$. The number of customers in the queue will tend to grow; what can we say about N_t, the queue-size at time t? If N_0 is not too small and if ρ is appreciably greater than unity, the chance is very small that the server will at any time be free in the interval $(0, t)$. In other words the server is busy serving customers throughout the interval. Therefore the number of customers served in time t, M_t say, is the maximum number of service-times that can be added together forming a total of t or less. The properties of M_t are considered in *renewal theory*. Simple approximate formulae for the mean and variance of M_t are available, namely

$$ EM_t \sim \frac{t}{b}, \qquad \text{var}(M_t) \sim \frac{C^2 t}{b}. $$

These are valid asymptotically for large t, but in practice usually give good approximations whenever $t \gg b/C^2$; the formulae are exact for all t if the distribution of service-time is exponential.

The number of arrivals in the same time period $(0, t)$ is L_t say, and we have, again from renewal theory, that

$$ EL_t \sim \frac{t}{a}; $$

we have already agreed to write $V(t)$ for the variance of L_t.

Now, obviously,

$$N_t = N_0 + L_t - M_t, \tag{9}$$

so that, on taking expectations,

$$\left.\begin{aligned} EN_t &\sim N_0 + t/a - t/b \\ &= N_0 + t(\rho - 1)/b. \end{aligned}\right\} \tag{10}$$

Further, if L_t and M_t are uncorrelated random variables, then (9) gives also that

$$\left.\begin{aligned} \text{var}(N_t) &= \text{var}(L_t) + \text{var}(M_t) \\ &\sim V(t) + C^2 t/b. \end{aligned}\right\} \tag{11}$$

Plainly, if N_0 has a probability distribution instead of being a fixed number as we have supposed so far, then var(N_0) must be added to the right-hand side of (11), and N_0 must be replaced by EN_0 in (10).

It is easy to obtain from (10) the expected queueing time, Q_t say, of a customer arriving at time t. He will have before him N_t customers in the queue, and if we neglect the fact that the leading customer may be partly served, Q_t is simply the sum of N_t service-times. Therefore

$$\left.\begin{aligned} EQ_t &\sim bEN_t \\ &\sim N_0 b + t(\rho - 1). \end{aligned}\right\} \tag{12}$$

Consider now the problem which is, in a sense, the converse to the rush-hour one. Suppose N_0 is very large and that ρ is fairly small and certainly less than one. Can we obtain any approximate results about N_t? Roughly speaking, the effects of the initial high value N_0 will be negligible when N_t has been reduced to the neighbourhood of $EN_\infty = \bar{N}$, the equilibrium mean of N_t for the given value of ρ. While N_t is declining to $\bar{N}$, the server will be constantly busy, so that we can employ the same kind of reasoning that we used for the rush-hour situation. If T is the time after which N_t has reduced to the value $\bar{N}$, then we have approximately

$$\bar{N} = N_0 + L_T - M_T. \tag{13}$$

But it is intuitively clear, and may be deduced from renewal theory, as were the earlier results used in deriving (10) and (11), that

$$EL_T \sim ET/a, \qquad EM_T \sim ET/b.$$

Hence we have from (13) that

$$\left. \begin{array}{l} ET \sim [EN_0 - \bar{N}]/[1/b - 1/a] \\[2mm] \quad = \dfrac{b(EN_0 - \bar{N})}{1 - \rho}. \end{array} \right\} \qquad (14)$$

The approximate results (10), (11) and (14) are admittedly based on rather rough-and-ready arguments. Nevertheless they are useful in that they provide rough guidance in situations where exact calculations are prohibitive.

We close this section with a numerical example. Suppose that we have a single-server queue with random arrivals at the rate of one a minute. Suppose service-times are constant and 45 seconds in duration. Then $\rho = \frac{3}{4}$ and (2.20) shows that $\bar{N} = \frac{15}{8}$. Now suppose that at zero time the arrival rate is suddenly doubled, so that $\rho = 1\frac{1}{2}$. Thus we can take $EN_0 = \frac{15}{8}$, and, supposing t measured in minutes, we have from (10) that

$$EN_t \simeq \tfrac{15}{8} + \tfrac{2}{3}t.$$

Hence, if the 'rush' period lasts one hour the expected size of the queue at its end is approximately 42 (actually the formula gives $41\frac{7}{8}$, but the approximations in our argument are undeserving of such numerical accuracy). To calculate the time that must further elapse for the system to return to its more usual condition, once the arrival rate drops to one a minute at the end of the 'rush-hour', we employ (14). We write in this formula $EN_0 = 42$, $\bar{N} = 2$ (as an approximation to the more correct $\frac{15}{8}$), $b = \frac{3}{4}$, $\rho = \frac{3}{4}$. Then $ET = 120$, i.e. the increased arrival rate over a period of one hour builds up, on the average, such a large queue that two further hours are needed at the normal lower arrival rate to reduce the queue-size to its more usual proportions. Variances, and hence rough confidence intervals, corresponding to the predicted values of 42 and 2 hours, can be obtained from the formulae we have given.

Equation (14) can sometimes be used to estimate roughly whether the equilibrium probability distribution is likely to give a good approximation to the average properties of a system over a finite time t_0. To do this, suppose that the initial queue-size must exceed the equilibrium mean queue-size.

If

(*a*) the expected time (14) to go from the initial queue-size to the equilibrium average queue-size is very small compared with t_0, and

(*b*) the variance of the total time taken to serve all customers arriving in t_0 is large compared with the mean service-time,

then it is reasonable to use the equilibrium theory. If the initial queue-size is less than the equilibrium mean, (*a*) may be omitted. Of course, these conditions are no substitute for proper quantitative investigation of the non-equilibrium theory, but serve as a rough rule for use when detailed investigation, mathematically or by sampling experiments, cannot be undertaken.

3.2 Queues with many servers

In section 3.1 we dealt briefly with the non-equilibrium theory of the simplest queueing system. We now return to the equilibrium theory and consider a different type of complication of the simplest system: let there be $m > 1$ servers instead of the single one that we have usually assumed up to now.

Section 2.4 (iii) gives the equilibrium distribution of queue-size for a general value of m when arrivals are random and service-times exponentially distributed. For more general arrival patterns and service-time distributions the study of the m-server queue usually presents considerable mathematical difficulties. We shall deal here with one case where results can, however, be obtained without too much analytical complexity.

Suppose that intervals between successive arrivals are independently distributed with distribution function $A(x)$, and that service-times are independently exponentially distributed with density function $\sigma e^{-\sigma x}$. It is assumed that $A(x)$ is such that there is zero probability that two or more customers arrive simultaneously.

At the initial time let there be q_0 customers in the system, including any being served. Let the next customer to arrive find q_1 customers ahead of him, the next q_2, and so on. Thus q_n is the number of customers in the system immediately before the arrival of the nth customer, customers being numbered in order of arrival. Immediately after the nth customer arrives, the number of customers in the system is of course $q_n + 1$.

It is convenient next to define a measure of traffic intensity. If the m servers were busy all the time they would serve $m\sigma$ customers per unit time in the long run. In fact, customers arrive at average rate $1/a_1$ per unit time, where a_1 is the mean of the distribution $A(x)$. This leads us to define the traffic intensity ρ as $(m\sigma a_1)^{-1}$.

If $\rho < 1$ customers tend to arrive more slowly than they can be served and therefore it is reasonable to expect that the probability distribution of q_n will tend, as n increases, to a stable or equilibrium distribution, i.e. that

$$\lim_{n \to \infty} \text{prob}(q_n = k) = p_k, \text{ say.}$$

The existence of p_k can be proved rigorously, but here we shall take the existence for granted and concentrate on the explicit calculation of p_k; it will be assumed throughout that $\rho < 1$.

It is important to note that the $\{q_n\}$ here are not the same as the $\{q_n\}$ of section 2.6; there the queue-size was considered at the instants when service is just completed, whereas now we deal with instants at which arrivals occur. The general idea of studying the process at discrete time points only is the same in both cases, but it is essential, in applying the method, to choose the defining instants appropriately.

Let

$$p_{ij} = \text{prob}(q_{n+1} = j \mid q_n = i).$$

F

Then p_{ij} is the probability that exactly $i+1-j$ customers are served during an arrival interval; consequently $p_{ij} = 0$, if $j > i+1$. If $j \geqslant m$, all servers will have been busy throughout the arrival interval concerned. Therefore there is, for any time period δt in the arrival interval, a probability $m\sigma\delta t$ that the service of some customer is completed. It follows, by the argument leading to (2.16), that

$$p_{ij} = \int\limits_{0}^{\infty} \frac{e^{-m\sigma x}(m\sigma x)^{i-j+1}}{(i-j+1)!} \, dA(x) \quad (m \leqslant j \leqslant i+1); \quad (15)$$

note that when $j \geqslant m$, p_{ij} depends only on $i-j$ and to emphasize this we shall write $p_{ij} = \varpi_{i-j+1}$.

To picture the development of the sequence $\{q_n\}$ consider a particle moving between 'states' E_0, E_1, ... If say $q_n = k$, the particle goes to E_k at the nth move. The successive locations of the particle will, of course, be governed by the transition probabilities, p_{ij}, which we have already introduced. Let us place the particle in the state corresponding to the initial queue-size q_0 and watch its progress over a very large number t of moves. From time to time E_k will occur; write $N_k(t)$ for the total number of such occurrences. After an occurrence of E_k two things can happen: (i) the particle moves to a state E_j with $j \leqslant k$ (this includes the possibility that E_k may occur again immediately); (ii) the particle moves to E_{k+1}. If (ii) occurs, the particle will, in general, move about for a while among the states E_j with $j > k$ until it moves to a state E_j with $j \leqslant k$. Once it has moved to E_j with $j \leqslant k$, however, then E_k must occur before E_{k+1} can occur again. This is because the particle can move upwards only one state at a time, although it can of course jump in one move to any state with a lower suffix. We shall denote by λ_k the mean number of occurrences of E_{k+1} between successive occurrences of E_k.

A very important property of this system is that λ_k is constant, equal to λ say, for all $k \geqslant m-1$. This may be seen as follows. The probability that the particle moves directly to E_{k+1} from E_k is ϖ_0, for all $k \geqslant m-1$; thus the probability of zero occurrences of E_{k+1}

between two occurrences of E_k is $1 - \varpi_0$ for all $k \geqslant m - 1$. On the other hand, if we place the particle in E_{k+1}, then the probability γ, say, that it returns to E_{k+1} before coming to a state E_j with $j \leqslant k$ cannot depend on k. This is because, for all transitions like $E_i \to E_j$, with $j \geqslant m$, the transition probabilities depend only on $i - j$. Thus the probability of exactly r occurrences of E_{k+1} between two occurrences of E_k is $\varpi_0 \gamma^{r-1}(1 - \gamma)$, and hence $\lambda_k = \varpi_0/(1 - \gamma)$ for all $k \geqslant m - 1$.

Thus to every occurrence of E_k $(k \geqslant m - 1)$ we can expect on the average λ occurrences of E_{k+1}. Therefore as t becomes very large, $N_{k+1}(t)/N_k(t)$ tends to λ. But the equilibrium probability p_k is the proportion of a very long time spent in E_k (section 2.2); hence $p_{k+1} = \lambda p_k$ $(k \geqslant m - 1)$, and therefore

$$p_k = C\lambda^k \quad (k \geqslant m - 1),$$

where C is a constant. Thus we have established by quite simple arguments that the distribution of queue-size is geometric, possibly modified in the first m terms. We have, however, still to determine the constants C and λ.

The calculation of λ is not too difficult. We have that

$$\text{prob}(q_{n+1} = m) = \sum_{j=m-1}^{\infty} \text{prob}(q_n = j \text{ and } q_{n+1} = m).$$

from which

$$p_m = \sum_{j=m-1}^{\infty} p_j p_{jm}$$

or

$$C\lambda^m = \sum_{j=m-1}^{\infty} C\lambda^j \varpi_{j-m+1}$$

or

$$\lambda = \sum_{r=0}^{\infty} \lambda^r \varpi_r.$$

If now we use equation (15) for $\varpi_r = p_{r+m-1, m}$, we obtain for λ the transcendental equation

$$\lambda = A^{\star}(m\sigma - m\sigma\lambda). \tag{16}$$

It can be shown that when $\rho < 1$ there is a unique real root of (16) in the range $0 < \lambda < 1$; it is this root that gives the geometric ratio in the distribution of queue-size. This will not be proved here but instead two simple examples will be given.

Example (i). *Two servers and random arrivals.* If arrivals are at random with rate α, then $A^\star(s) = \alpha/(\alpha+s)$ and (16) becomes

$$\lambda = \frac{\alpha}{\alpha + 2\sigma - 2\sigma\lambda}.$$

This is a quadratic equation in λ whose roots are 1 and $\frac{1}{2}\alpha/\sigma = \rho$. Thus the equilibrium distribution of queue-size is of the form

$$p_k = C\rho^k \quad (k \geqslant 1).$$

This agrees with the result of section 2.4 (iii).

Example (ii). *Ten servers and regular arrivals.* Here $A^\star(s) = e^{-a_1 s}$ and (16) is equivalent to

$$\log \lambda = -10\sigma a_1(1-\lambda),$$

i.e., $\qquad\qquad \log \lambda = -(1-\lambda)/\rho.$

This can be solved numerically given the value of ρ. Thus with $\rho = \frac{1}{2}$, $\lambda = 0.203$, and for $k \geqslant 9$, $p_k = C(0.203)^k$, for some constant C.

If a customer arrives to find fewer than m customers in the system, then that customer will not have to wait for service, since there will be at least one server free to deal with him. Thus the probability that a customer will have to queue is

$$\sum_{k=m}^{\infty} p_k = \sum_{k=m}^{\infty} C\lambda^k = \frac{C\lambda^m}{1-\lambda}.$$

Therefore, given that a customer will have to wait, the conditional probability that he will find a queue of length k is

$$C\lambda^k \bigg/ \left\{\frac{C\lambda^m}{1-\lambda}\right\} = \lambda^{k-m}(1-\lambda) \quad (k \geqslant m).$$

Thus the distribution of the number of customers actually queue-ing, conditionally upon there being any at all, is geometric with common ratio λ for any many-server queue with general independent arrivals and exponential service-times.

We can obtain from this result the conditional distribution of queueing-time for a customer, given that his queueing-time is non-zero. For this we assume that the queue-discipline is 'first-come, first-served'. Suppose that on arrival the customer finds $m + l$ customers ahead of him in the system. Then he will have to queue until $l + 1$ service operations have been completed; such completions occur in a Poisson process at rate $m\sigma$, since there are throughout m customers being served. Therefore, with probability $\lambda^l(1 - \lambda)$, the queueing-time has the distribution of the sum of $l + 1$ exponentially distributed random variables. The determination of the probability density function of queueing-time now follows exactly the argument leading to (2.30, 2.31); the answer is

$$m\sigma(1 - \lambda) \exp[-m\sigma(1 - \lambda)x]. \tag{17}$$

That is, conditionally on being non-zero, queueing-time has an exponential distribution with mean $[m\sigma(1 - \lambda)]^{-1}$. It is remarkable that the exponential form of the distribution holds for all values of m and for all forms of the distribution, $A(x)$, of the intervals between successive arrivals.

We have obtained a certain amount of information about the system without evaluating the constant C. However the calculation of C is necessary for a full solution; for example the probability that a customer does not have to queue is

$$1 - \frac{C\lambda^m}{1 - \lambda}. \tag{18}$$

When $m = 1$ the calculation of C is trivial; for $p_k = C\lambda^k$ ($k \geq 0$), and the condition that $\sum p_k = 1$ leads immediately to the equation $C = 1 - \lambda$.

When $m > 1$, so that we have a many-server system, the calculation is more difficult. We shall indicate here how C can be determined when $m = 2$; results for larger values of m can be obtained by similar methods, or from the work of D.G. Kendall.*

Suppose that a customer arrives in the two-server queue and has to queue for service. What is the probability that the next customer to arrive will find both servers free? For this, the interval of time separating the two arrivals must be long enough for the queueing-time of the first customer to end and then for each server to finish serving the remaining customer before him. Write x for the arrival-interval under discussion, and let y be the queueing-time of the first customers; then y has the exponential distribution (17). The probability that in the time $x - y$ each server will serve the customer before him is easily found to be $[1 - e^{-\sigma(x-y)}]^2$. Thus, conditionally upon x, the required probability is

$$2\sigma(1-\lambda) \int\limits_0^x e^{-2(1-\lambda)\sigma y} [1 - e^{-\sigma(x-y)}]^2 \, dy,$$

which simplifies to

$$1 - \frac{4(1-\lambda)e^{-\sigma x}}{1-2\lambda} - \frac{(1-\lambda)e^{-2\sigma x}}{\lambda} + \frac{e^{-2(1-\lambda)\sigma x}}{\lambda(1-2\lambda)}.$$

If we now grant x the arrival distribution $A(x)$ and find the expectation of this last expression, we are led, on using (16), to

$$s = \frac{1-\lambda}{1-2\lambda} \left[2 - 4A^\star(\sigma) - \frac{1-2\lambda}{\lambda} A^\star(2\sigma) \right], \tag{19}$$

as the probability that a customer, who follows one obliged to queue, will find both servers free.

If a customer arrives to find one server busy and one free, the probability that the next customer to arrive will find both servers free is

$$p_{10} = \int\limits_0^\infty (1 - e^{-\sigma x})^2 \, dA(x)$$

$$= 1 - 2A^\star(\sigma) + A^\star(2\sigma). \tag{20}$$

* D. G. Kendall, *Ann. Math. Statist.*, **24** (1953), 338.

Further, given that a customer arrives to find both servers free, the probability that the next customer to arrive also finds both servers free is clearly

$$p_{00} = 1 - A^{\star}(\sigma). \tag{21}$$

We now apply these preliminary results in the following equation, whose correctness is simple to verify:

$$p_0 = p_0 p_{00} + p_1 p_{10} + rs, \tag{22}$$

where r is the probability that a customer will have to queue, and s is the probability that a customer who has to queue is followed by one who finds both servers free. In conjunction with (22) we use the known results that $p_k = C\lambda^k$ for $k \geq 1$, that $\sum p_k = 1$, that $1 - r = p_0 + p_1$, and equations (19), (20), (21). Routine manipulation then yields the results:

$$p_0 = \frac{(1-\lambda)[1 - 2A^{\star}(\sigma)]}{1 - \lambda - A^{\star}(\sigma)}, \tag{23}$$

$$C = \frac{(1-\lambda)(1-2\lambda)A^{\star}(\sigma)}{\lambda[1 - \lambda - A^{\star}(\sigma)]}. \tag{24}$$

The complete equilibrium distribution of queue-size is thus determined by (23) and (24), in terms of known quantities. Note that the probability that a customer actually queues is equal to

$$\frac{\lambda(1-2\lambda)A^{\star}(\sigma)}{1 - \lambda - A^{\star}(\sigma)}. \tag{25}$$

To complete our discussion of the general two-server queue, we apply our results to some specific examples.

Example (iii). Two-server queue with random arrivals and exponential service-times. This is a continuation of Example (i), in which we found that $\lambda = \rho$.

We also find that $A^{\star}(\sigma) = \alpha/(\alpha + \sigma) = 2\rho/(1 + 2\rho)$.

Thus, by (25), the probability that a customer will have to queue is

$$\frac{2\rho^2}{1+\rho}$$

and, by (23) and (24),

$$p_0 = \frac{1-\rho}{1+\rho}$$

$$p_k = \frac{2(1-\rho)\rho^k}{1+\rho} \quad (k \geqslant 1).$$

Example (iv). *Two-server queue with regular arrivals and exponential service-times.* Here $A^\star(s) = e^{-a_1 s}$ and, as in Example (ii), we are led to the following equation for λ:

$$-\rho \log \lambda = 1 - \lambda.$$

Once λ is determined, we have, by (25), that the probability a customer must queue is

$$\frac{\lambda(1-2\lambda) e^{-\frac{1}{2}\rho^{-1}}}{1-\lambda-e^{-\frac{1}{2}\rho^{-1}}}.$$

Also, from (23) and (24), we have that

$$p_0 = \frac{(1-\lambda)(1-2e^{-\frac{1}{2}\rho^{-1}})}{1-\lambda-e^{-\frac{1}{2}\rho^{-1}}},$$

$$p_k = \frac{(1-\lambda)(1-2\lambda) e^{-\frac{1}{2}\rho^{-1}} \lambda^k}{\lambda(1-\lambda-e^{-\frac{1}{2}\rho^{-1}})} \quad (k \geqslant 1).$$

3.3 Queues with priorities

(i) *Introduction*

Sometimes the queue-discipline may be such that some types of customer receive priority. For example, the cost per unit time of keeping certain customers queueing may be particularly high and it may then be reasonable to give them priority. Or, if the cost per unit queueing-time is constant, it will be desirable to reduce the overall mean queueing-time, and this can be achieved by giving high priority to customers expected to have a small service-time.

It is simplest mathematically to deal with a priority system in which a customer, once at the service-point, remains there until his service is complete. Then the next customer for service is the one of highest priority among those queueing. We call this non-preemptive priority. A preemptive priority system is one in which a customer of high priority takes, on arrival, immediate precedence over customers of lower priority, the customer whose service is interrupted returning to the service-point only when there are no higher priority customers remaining in the system. Numerous intermediate and more complicated priority rules are possible.

The first and major part of this section is devoted to non-preemptive priority and the second part to preemptive priority.

(ii) *Non-preemptive priority*

Suppose that each customer has a priority class $1, 2, \ldots, k$, where 1 is the highest priority and k the lowest. A customer of class j will be called for short, a j-customer, and has non-preemptive priority over customers of class $j+1, \ldots, k$; customers of a given class are served in order of arrival. Assume that customers of different classes arrive independently at random at rates $\alpha_1, \ldots, \alpha_k$ and that the unit of time is chosen so that the total arrival rate

$$\alpha_1 + \ldots + \alpha_k = 1.$$

Then α_j is the probability that a particular arrival is a j-customer. Let the service-times of different customers be independently distributed with $B_j(x)$ the distribution function for j-customers. Finally we shall consider only a single-server system, although some of the results can be extended to several-server systems.

The 'overall' service-time distribution is

$$B(x) = \sum_{j=1}^{k} \alpha_j B_j(x).$$

For convenience we shall, *in this section only*, denote by b_j, c_j, ... the first, second, ... moments of $B_j(x)$. For instance

$$c_j = \int\limits_0^\infty x^2 \, dB_j(x).$$

Similarly b, c, ... will be used for the moments of $B(x)$, so that

$$b = \sum \alpha_j b_j, \qquad c = \sum \alpha_j c_j.$$

Further, since $\sum \alpha_j = 1$, the traffic intensity ρ is equal to b.

We shall suppose throughout that $\rho < 1$ and shall assume, without further discussion, that the properties of the system have a stationary probability distribution. Our object is to study some properties, of practical interest, of this stationary distribution.

The argument to be used depends on the device, used previously in section 2.6, of considering the system only at those instants, to be called *epochs*, at which a service operation is just completed. If at such an epoch, the customer at the head of the queue, i.e. whose service is about to start, is a j-customer we shall call it a j-epoch and say that the event R_j occurs. If at any epoch there are no customers in the queue, we call it a 0-epoch and say that R_0 occurs. Let π_j be the stationary probability of R_j. One of the reasons for considering the epochs R_j is that the queueing-time of the leading j-customer has just ended, and the number of j-customers in the queue behind him is the number who arrived during that queueing-time.

We can simplify the mathematical problem appreciably by the following device. Fix an integer j greater than 2 and consider a modified system in which customers of class 1, 2, ..., $j-1$ are combined into a single priority class H. Customers in H in the modified system lose their priority classification and are served in order of arrival. However, H-customers continue to have priority over customers of class $j, j+1, ..., k$. In the modified system events $R_1, ..., R_{j-1}$ no longer arise; instead we have an event R_H, which occurs at any epoch at which there is an H-customer in the queue. Now events R_H in the modified system do not in general occur at the same instants as events $R_1, \ldots, R_{j-1}$ in the original system, because the order in which the customers are served is different in the two cases. What is important to us, however, is that the j-epochs occur at the same instants in the original and modified systems, and further that the distribution of the number of j-customers in the queue at a j-epoch is the same in both systems. The reason for this is that a j-epoch can occur only when no H-customers remain in the system, and the periods during which the server is busy serving H-customers may be seen to be identical in the original and modified systems.

The mathematical significance of this result is that in order to obtain results for j-customers it is enough to take $j = 2$. For example we may quote a formula to be proved later in the section: the expected number of 2-customers in the queue at a 2-epoch is

$$1 + \frac{\frac{1}{2}\alpha_2 c}{\rho(1 - \alpha_1 b_1)(1 - \alpha_1 b_1 - \alpha_2 b_2)}. \tag{26}$$

If we let $\alpha_1 \to 0$, we have a result for the situation in which top-priority customers never arrive, i.e. in which 2-customers have effectively top priority. Thus the expected number of 1-customers in the queue at 1-epochs is

$$1 + \frac{\frac{1}{2}\alpha_1 c}{\rho(1 - \alpha_1 b_1)}.$$

Consider now j-customers, $j > 2$. Combine customers of class $1, \ldots, j-1$ into a single class H, thereby making j-customers effectively 2-customers. Therefore, by (26), the required expected number is

$$1 + \frac{\frac{1}{2}\alpha_j c}{\rho(1 - \alpha_H b_H)(1 - \alpha_H b_H - \alpha_j b_j)},$$

where α_H, b_H refer to the new class H of customers and ρ and c are unaffected by the combination of priority classes. It is easy to show that

$$\alpha_H b_H = \alpha_1 b_1 + \ldots + \alpha_{j-1} b_{j-1},$$

so that the expected number of j-customers at a j-epoch is

$$1 + \frac{\frac{1}{2}\alpha_j c}{\rho\left(1 - \sum_{i=1}^{j-1} \alpha_i b_i\right)\left(1 - \sum_{i=1}^{j} \alpha_i b_i\right)}. \tag{27}$$

Thus the formula (26) for 2-customers is easily generalized to j-customers for arbitrary j.

If we combine all customers with one class H, the earlier argument shows that the probability π_0 of R_0 is unchanged. But the modified system now has no priorities and, with random arrivals, we have by (2.19) that $\pi_0 = 1 - \rho$. Further, of the arrivals in a very long time-period, a proportion α_j are j-customers. The proportion of j-epochs in this time-period is, except for end-effects, the proportion of those j-customers that arrived to find the system not empty, and this proportion is $\rho\alpha_j$; this is because any customer, of whatever class, has a probability $1 - \rho$ of arriving to find the system empty. Therefore

$$\pi_j = \rho\alpha_j \quad (j = 1, 2, \ldots, k). \tag{28}$$

Consider now the calculation of the stationary distribution of the number of 2-customers at a 2-epoch. Let T be an arbitrary epoch and T' the following epoch. Let q_1, q_2 be the numbers of 1- and 2-customers in the queue at T and q_1', q_2' the corresponding numbers at T'. We proceed by relating the pair (q_1', q_2') to the pair (q_1, q_2). There are a number of cases to consider depending on the class of the epoch T.

Suppose first that T is a j-epoch for some $j = 1, 2, \ldots, k$. Then the interval from T to T' is the service-time of a j-customer. Let ξ_{j1}, ξ_{j2} be the numbers of a 1- and 2-customers arriving during the period (T, T'). The joint distribution of ξ_{j1}, ξ_{j2} can be specified by their probability generating function. For given S_j, the service-time of the customer entering the service-point at T, ξ_{j1} and ξ_{j2} are independently distributed in Poisson distributions with means $\alpha_1 S_j$ and $\alpha_2 S_j$. Thus the conditional generating function is

$$E(z_1^{\xi_{j1}} z_2^{\xi_{j1}} \mid S_j) = \exp\left[-\alpha_1 S_j(1 - z_1) - \alpha_2 S_j(1 - z_2)\right].$$

Therefore the unconditional generating function, obtained by integrating with respect to the distribution of S_j, is

$$E(z_1^{\xi_{j1}} z_2^{\xi_{j2}}) = \int_0^\infty E(z_1^{\xi_{j1}} z_2^{\xi_{j2}} \mid x) \, dB_j(x)$$

$$= B_j^\star[\alpha_1(1-z_1) + \alpha_2(1-z_2)]$$

$$= B_j^\star(u_{12}), \text{ say.}$$

Now if T is a 1-epoch

$$q_1' = q_1 + \xi_{11} - 1, \qquad q_2' = q_2 + \xi_{12}, \tag{29}$$

by an argument similar to that of section 2.6. If T is a 2-epoch,

$$q_1' = \xi_{21}, \qquad q_2' = q_2 + \xi_{22} - 1, \tag{30}$$

and if T is a j-epoch $(2 < j \leqslant k)$

$$q_1' = \xi_{j1}, \qquad q_2' = \xi_{j2}. \tag{31}$$

Finally, if T is a 0-epoch, the server becomes free at T and with probability α_j the next customer to arrive is of class j. Thus, with probability α_j,

$$q_1' = \xi_{j1}, \qquad q_2' = \xi_{j2}. \tag{32}$$

If we combine these formulae for (q_1', q_2') and take generating functions, we have that

$$E(z_1^{q_1'} z_2^{q_2'}) = \pi_1 E(z_1^{q_1 + \xi_{11} - 1} z_2^{q_2 + \xi_{11}} \mid R_1 \text{ occurs at } T)$$

$$+ \pi_2 E(z_1^{\xi_{11}} z_2^{q_2 + \xi_{22} - 1} \mid R_2 \text{ occurs at } T)$$

$$+ \sum_{j=3}^k \pi_j E(z_1^{\xi_{j1}} z_2^{\xi_{j2}} \mid R_j \text{ occurs at } T)$$

$$+ \pi_0 \sum_{j=1}^k \alpha_j E(z_1^{\xi_{j1}} z_2^{\xi_{j2}} \mid R_j \text{ occurs at } T).$$

If we use the independence of the ξ's and the q's and write

$$G_1(z_1, z_2) = E(z_1^{q_1} z_2^{q_1} \mid R_1 \text{ occurs at } T),$$
$$G_2(z_2) = E(z_2^{q_2} \mid R_2 \text{ occurs at } T),$$

we obtain the equation

$$\left. \begin{aligned} E(z_1^{q_1} z_2^{q_1}) &= \frac{\pi_1}{z_1} B_1^\star(u_{12}) G_1(z_1, z_2) + \frac{\pi_2}{z_2} B_2^\star(u_{12}) G_2(z_2) \\ &\quad + \sum_{j=3}^{k} \pi_j B_j^\star(u_{12}) + \pi_0 \sum_{j=1}^{k} \alpha_j B_j^\star(u_{12}). \end{aligned} \right\} \quad (33)$$

The final step is to introduce the stationarity of the system, according to which (q_1', q_2') has the same distribution as (q_1, q_2). Thus

$$E(z_1^{q_1} z_2^{q_2}) = \pi_1 G_1(z_1, z_2) + \pi_2 G_2(z_2) + 1 - \pi_1 - \pi_2. \quad (34)$$

Therefore, combining (33) and (34), we have that

$$\pi_1 G_1(z_1, z_2)[1 - B_1^\star(u_{12})/z_1] + \pi_2 G_2(z_2)[1 - B_2^\star(u_{12})/z_2]$$
$$+ \sum_{j=3}^{k} \pi_j[1 - B_j^\star(u_{12})] + \pi_0[1 - \sum_{j=1}^{k} \alpha_j B_j^\star(u_{12})] = 0, \quad (35)$$

where it will be recalled that $u_{12} = \alpha_1(1 - z_1) + \alpha_2(1 - z_2)$.

This rather unpromising-looking equation does, in fact, give the information we want. First, if we differentiate partially with respect to z_i and put $z_1 = z_2 = 1$, we get two equations which, after manipulation, lead to an alternative proof of the result (28), $\pi_j = \rho \alpha_j$.

We can obtain similarly three equations by setting $z_1 = z_2 = 1$ after taking second derivatives with respect to z_1 and z_2 and these can be reduced after rather heavy algebra to the following:

$$(1 - \alpha_1 b_1) m_{11} = 1 - \alpha_1 b_1 + (2\rho)^{-1} \alpha_1 c, \quad (36)$$

$$(1 - \alpha_2 b_2) m_{22} = 1 - \alpha_2 b_2 + \alpha_1 b_1 m_{12} + (2\rho)^{-1} \alpha_2 c, \quad (37)$$

$$(1 - \alpha_1 b_1) m_{12}/\alpha_2 = b_1 m_{11} + b_2 m_{22} + \rho^{-1} c - b_1 - b_2, \quad (38)$$

where

$$m_{11} = \frac{\partial}{\partial z_1} G_1(z_1, z_2), \qquad m_{12} = \frac{\partial}{\partial z_2} G_1(z_1, z_2), \qquad m_{22} = \frac{\partial G_2(z_2)}{\partial z_2},$$

all derivatives being evaluated at $z_1 = z_2 = 1$. The m's have simple interpretations and m_{22} is the one we want, for, from the definition of $G_2(z_2)$,

$$m_{22} = E(q_2 \mid R_2 \text{ occurs at } T),$$

i.e. m_{22} is the expected number of 2-customers in the queue at a 2-epoch. On solving (36), (37), (38), we have that

$$m_{22} = 1 + \frac{\alpha_2 c}{2\rho(1 - \alpha_1 b_1)(1 - \alpha_1 b_1 - \alpha_2 b_2)}, \qquad (39)$$

the result earlier generalized to arbitrary j in (27). One fewer than this is the expected number of j-customers to arrive during the queueing-time of the leading j-customer. Since these customers arrive randomly at rate α_j, it follows that Q_j, the mean queueing-time of a j-customer, is, on allowing for the chance $1 - \rho$ of not having to queue,

$$Q_j = \frac{\frac{1}{2}c}{\left(1 - \sum_{i=1}^{j-1} \alpha_i b_i\right)\left(1 - \sum_{i=1}^{j} \alpha_i b_i\right)} \qquad (40)$$

The mean queueing-time, Q, of all customers is

$$Q = \sum \alpha_j Q_j$$

$$= \frac{c}{2} \sum_{j=1}^{k} \frac{\alpha_j}{\left(1 - \sum_{i=1}^{j-1} \alpha_i b_i\right)\left(1 - \sum_{i=1}^{j} \alpha_i b_i\right)} \qquad (41)$$

Equations (40) and (41) are the key formulae, from which the effect on mean queueing-time of any proposed system of priorities can be assessed.

We now consider briefly the choice of a strategy for assigning priorities. Suppose that there are k types of customer and that the cost of keeping a customer of the jth type queueing for unit time is constant and equal to w_j. This implies that the cost of queueing depends only on the mean queueing-time and that the mean cost is

$$C = \sum \alpha_j w_j Q_j$$

$$= \frac{c}{2} \sum_{j=1}^{k} \frac{\alpha_j w_j}{\left(1 - \sum_{i=1}^{j-1} \alpha_i b_i\right)\left(1 - \sum_{i=1}^{j} \alpha_i b_i\right)}. \tag{42}$$

Formula (42) holds when the ordering of priorities is $(1, 2, \ldots, k)$; we are interested now in choosing priorities to minimize (42). Suppose that $k > 3$ and that in (42) we change all 2's into 3's and vice versa. This gives us a new mean cost C' which would apply if the classes 2 and 3 had their priorities interchanged. Only two terms in (42) are affected and after some algebra we find that

$$C - C' = \frac{c\Delta}{2}\left(\frac{w_2}{b_2} - \frac{w_3}{b_3}\right),$$

where

$$\Delta = (1 - \alpha_1 b_1 - \alpha_2 b_2)^{-1} + (1 - \alpha_1 b_1 - \alpha_3 b_3)^{-1}$$
$$- (1 - \alpha_1 b_1)^{-1} - (1 - \alpha_1 b_1 - \alpha_2 b_2 - \alpha_3 b_3)^{-1}.$$

Now, if x, y, z are positive, and different, it may be verified that

$$\frac{1}{x-y} + \frac{1}{x-z} - \frac{1}{x} - \frac{1}{x-y-z} < 0.$$

Hence $\Delta < 0$, so that $C < C'$ if and only if $b_2/w_2 < b_3/w_3$; thus if $b_2/w_2 > b_3/w_3$ we can reduce the mean cost by changing priority classifications of 2-customers and 3-customers. The same argument, with slight changes in the algebra, shows that if

$$b_{j-1}/w_{j-1} > b_j/w_j,$$

we can reduce mean cost by changing the priority classification for the $(j-1)$-customer and the j-customers. This holds for

$2 < j < k$ and a slightly simpler argument deals with the extreme cases $j = 2, k$. Thus, by changing priority classifications, we can go on reducing mean cost until we arrive at a priority system based on the quantity

$$\frac{\text{mean service-time}}{\text{cost of queueing for unit time}},$$

low values corresponding to high priority. We then have a priority scheme of minimum mean cost. It is interesting that the optimum priority rule does not depend on the higher moments of the service-time distributions or on the proportions $\alpha_1, \ldots, \alpha_k$.

A special case is when the cost w_j is the same for all groups, when the optimum priority number of a class depends only on the mean service-time of the class. A limiting case of this arises when it is possible to predict the service-time of a customer accurately on arrival, and when service-time has a continuous distribution with p.d.f. $b(x)$. We can then imagine an infinite number of priority classes, the customer selected for service being, by what was proved above, always the one with lowest service-time. The resulting mean queueing-time is then easily shown from (41) to be

$$Q_{\min} = \frac{c}{2} \int_0^\infty \frac{b(x)\, dx}{\left[1 - \int_0^x y b(y)\, dy\right]^2}. \tag{43}$$

If the queue-discipline 'first come, first served' is used, the mean queueing-time is, from section 2.6

$$Q_0 = \frac{c\rho}{2(1-\rho)b},$$

and $Q_{\min}$ can be appreciably less than this.

The strategy based on service-time is optimum when the service-time of each customer can be predicted accurately on arrival and when all that is known about future customers is the random arrival rate $\alpha = 1$ and the p.d.f. of service-time. If something more is known, e.g. if the arrival instants and service-times

G

of the next few customers can be predicted, the strategy is not in general optimum, although it will often be good.

A simpler priority system based on the service-time is as follows. Suppose that $b(x) = \sigma e^{-\sigma x}$ where $\rho = \sigma^{-1}$, and that we count all customers whose service-times are less than or equal to $\phi\rho$ as 1-customers, all other customers being 2-customers. It can be shown that for this priority system

$$Q = \frac{c(1-\rho+\rho e^{-\phi})}{2(1-\rho)(1-\rho+\rho e^{-\phi}+\rho\phi e^{-\phi})}.$$

If the cost per unit queueing-time is the same for all customers, it is natural to choose ϕ to minimize Q, a matter that can be settled by elementary calculus. The optimum ϕ satisfies the equation

$$\frac{1}{\rho} = 1 + \frac{e^{-\phi}}{\phi-1}.$$

Fig. 3.1 shows the relation between ρ and ϕ. For example if $\rho = 0.4$, we should make the division at $1.2 \times$ mean service-time.

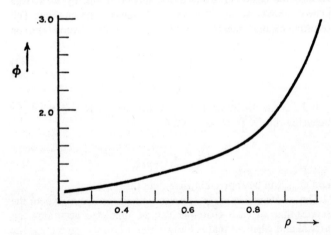

Fig. 3.1. Relation between traffic intensity ρ and parameter ϕ determining the optimum division into two priority classes.

Always $\phi > 1$ and as ρ approaches 1, ϕ becomes very large. It is possible also to find the reduction in mean queueing-time, as compared with 'first come, first served', attained by giving ϕ its optimum value. For example if $\rho = 0.75$, there will be a 37 per cent reduction in mean queueing-time by the introduction of this form of priority system; from Fig. 3.1 the system consists in giving high priority to customers with a service-time of less than $1.6 \times$ mean service-time, and low priority to other customers. Further gains will be possible if more than two priority classes are introduced. Of course, the introduction of a priority system of any form has no effect on the proportion of time for which the server is busy. This is determined solely by the total service-time of the customers arriving over a long period.

In a general way the more variable is the distribution of service-time, the greater will be the advantages of a priority system based on service-time.

We have worked so far only with mean queueing-time and this is all we need if the cost per unit queueing-time is constant for a given class of customers. It is possible in principle to obtain from (35) the higher moments of the number of 2-customers in the queue at 2-epochs and even $G_2(z_2)$, the probability generating function. For the moments we differentiate (35) repeatedly. An outline of the argument for finding $G_2(z_2)$ is as follows. If $0 < z_2 < 1$, there exists a unique z_1, depending on z_2, such that $0 < z_1 < 1$ and $B_1^\star(u_{12}) = z_1$. If we call this value of z_1, $\psi(z_2)$ and put $z_1 = \psi(z_2)$ in (35), then the first term vanishes and we obtain a formula for $G_2(z_2)$ as a function of z_2. Except in the simplest problems, however, this method is unlikely to be much use, because of the difficulty of finding $\psi(z_2)$.

(iii) *Preemptive priority*

We now consider more briefly a simple system with preemptive priority, i.e. in which a customer of low priority at the service-point is displaced by a customer of higher priority immediately the latter customer arrives. Suppose that there are just two classes of customer with independent random arrival rates α_1, α_2 and with service-times exponentially distributed with parameters σ_1, σ_2.

The time scale is again to be chosen so that $\alpha_1 + \alpha_2 = 1$. Several assumptions are possible concerning the total service-time of 2-customers who are interrupted at the service-point. The simplest, and the one we shall make here, is that after interruption service is resumed at the point where it was left off. Then, whenever a 2-customer is at the service-point, there is a constant probability $\sigma_2 \delta t + o(\delta t)$ of service being completed in $(t, t + \delta t)$. We suppose there to be one server, always available.

In any system with preemptive priority, properties referring only to 1-customers can be of course obtained by ignoring other customers. In particular, in the present problem, the equilibrium probability that there are m 1-customers in the system is, from (2.11),

$$\rho_1^m (1 - \rho_1) \quad (m = 0, 1, 2, \ldots),$$

where $\rho_1 = \alpha_1 / \sigma_1$, and the mean waiting time of a 1-customer is

$$\frac{1}{\sigma_1 (1 - \rho_1)}.$$

Let $p_{m,n}$ be the equilibrium probability that there are m 1-customers and n 2-customers in the system. The equilibrium equations are most neatly summarized in the form

$$\{\alpha_1 + \alpha_2 + \sigma_1 \epsilon(m) + \sigma_2 \epsilon(n)[1 - \epsilon(m)]\} p_{m,n}$$
$$= \alpha_1 p_{m-1,n} + \alpha_2 p_{m,n-1} + \sigma_1 p_{m+1,n}$$
$$+ \sigma_2 [1 - \epsilon(m)] p_{m,n+1} \quad (m, n = 0, 1, 2, \ldots), \quad (44)$$

where $\epsilon(m) = 1 \ (m \neq 0)$, $\epsilon(0) = 0$, and where it is understood that any p with a negative suffix is zero. The reader is advised to derive separately equations for the cases $m > 0$; $m = 0, n > 0$; $m = n = 0$, and to check that they fit the general form (44). For example on the right-hand-side the term in σ_2 is operative only when $m = 0$, for it is only then that a 2-customer is at the service-point.

The general solution of (44) is complicated. It can be shown that the mean number of 2-customers is

$$\sum_{m,n=0}^{\infty} n p_{m,n} = \frac{\rho_2}{1 - \rho_1 - \rho_2} \left[1 + \frac{\sigma_2 \rho_1}{\sigma_1 (1 - \rho_1)} \right].$$

Let $W_i^{(p)}$ be the mean waiting-time of i-customers; $W_1^{(p)}$ is given by (2.22). In a long time T there are $\alpha_i T$ arrivals of i-customers, spending in the system a total time $\alpha_i W_i^{(p)} T$. Therefore, the mean number of i-customers is $\alpha_i W_i^{(p)}$. Thus

$$W_2^{(p)} = \frac{1/\sigma_2}{1-\rho_1-\rho_2}\left[1+\frac{\sigma_2\rho_1}{\sigma_1(1-\rho_1)}\right]. \tag{45}$$

Thus the mean queueing-time of i-customers, $Q_i^{(p)}$, is given by

$$Q_1^{(p)} = \frac{\rho_1}{\sigma_1(1-\rho_1)}, \tag{46}$$

$$Q_2^{(p)} = \frac{1}{\sigma_2(1-\rho_1-\rho_2)}\left[\rho_1+\rho_2+\frac{\sigma_2\rho_1}{\sigma_1(1-\rho_1)}\right]. \tag{47}$$

The corresponding formulae for non-preemptive priority are, from special cases of (40),

$$Q_1 = \frac{\rho_1/\sigma_1+\rho_2/\sigma_2}{1-\rho_1},$$

$$Q_2 = \frac{\rho_1/\sigma_1+\rho_2/\sigma_2}{(1-\rho_1)(1-\rho_1-\rho_2)}$$

The mean queueing-time of all customers is $\alpha_1 Q_1 + \alpha_2 Q_2$.

In the system without priorities the mean queueing-time is, from (2.22),

$$\frac{\rho_1/\sigma_1+\rho_2/\sigma_2}{1-\rho_1-\rho_2}.$$

(iv) *Server availability*

One application of the formulae in sections (ii) and (iii) is to the approximate study of the effect on simple queueing of limited server availability. We suppose that for certain periods the server is unable to serve customers because, for example, of other duties, need for relaxation, etc. We think of these additional duties, etc., as the service of a second type of customer, possibly fictitious;

clearly there are many possible cases depending on which type of work has priority and of what type that priority is. In practice, the nature of the priorities may not be at all clearly defined and a semi-empirical argument may be necessary. There is more about this in section 4.3 (ii).

The treatment just suggested implies that there is an infinite reservoir of 'customers' producing additional work, and that queues of such customers can therefore arise. In some contexts, particularly when the additional work is all of one type, it may be more appropriate to suppose there to be a single 'customer', so that no further arrivals of that type can occur when a 'customer' is in the system.

CHAPTER IV

Machine Interference

4.1 Statement of problem

In this chapter we deal in rather more detail with a special queueing problem, that of machine interference. There are two reasons for our considering this problem in particular detail. It is one of considerable practical importance. It is also desirable to see how the purely theoretical considerations of the last two chapters may, by judicious modifications and assumptions, be made to apply to a practical situation. The complications that arise in connexion with the machine interference problem are typical of those in other queueing situations.

In the simplest form of the problem, one operative has charge of a number n of similar machines. From time to time a machine stops and does not resume production until it has been attended by the operative. If two or more machines are stopped at the same time only one can be attended at once, and so production is lost while machines are stopped awaiting attention. What happens should be clear from Fig. 4.1; at A a machine stops and is attended by the operative for the period AB. Meanwhile at C a second machine has stopped but has to wait until B before its period of attention begins.

Examples from engineering are when the machines are thread-cutting machines with up to twenty or more in the charge of one girl, and wire-drawing machines in small groups per operator. There are numerous instances in the textile industries notably in the spinning, winding and weaving sections. Another interesting example is the following: the 'machines' are men who from time to time have to stop their productive work in order, say, to grind their tools, and the 'operative' is the grinding machine. Only one man can grind his tools at a time so that production is lost while men wait their turn at the grinding machine. While this sounds different from the previous applications, a comparison with Fig. 4.1 shows that it is essentially the same situation.

The difference between this queueing situation and the ones with which we have been predominantly concerned in earlier sections is that here there is a finite population of customers (machines). Thus, if at some instant r machines are stopped, there are only $n - r$ machines running and the instantaneous arrival rate of new customers is proportional to this number, $n - r$.

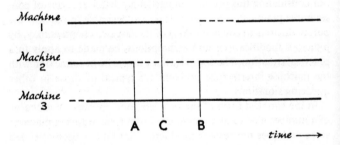

Fig. 4.1. Machine interference. Three machines under the care of one operative.

——— *Machine running.*

– – – *Machine stopped and under attention from operative.*

Production is lost not only while the machines are under attention by the operative but also while the machines are stopped awaiting attention (periods such as *CB* in Fig. 4.1). Such a loss is sometimes said to arise from *machine interference*. The rates of production per machine, and per operative, thus depend not only on the rate of occurrence of stops and on the 'attention-time' per stop, but also on the number, n, of machines per operative. Therefore it is natural to try to predict the rate of production from information about the stoppage-rates, the attention-times, and the number of machines per operative. There are several ways in which such a prediction can be useful in practice:

(a) We may wish to determine the optimum number of machines per operative on economic considerations, and to do this we must be able to work out the rate of production corresponding to any given n.

(b) We may contemplate a modification to the machines to eliminate certain stoppages; what should the new value of n be, and will the gain in production be sufficient to justify the modification?

(c) We may have measured rates of production from which to make a comparison to two alternative processes, but for one process there were say 10 machines per operative and for the other 15 machines per operative. Is an apparent difference between the processes due solely to the differing values of n?

To begin with we shall assume that the operative has no duties other than attending to stopped machines, i.e. that there is complete server availability. This is an important assumption to be discussed again later. We also assume that different machines have, in the long run, the same stoppage rates and attention times.

The first step is to define a dimensionless servicing factor λ, somewhat analogous to the traffic intensity in a simple queue. With one type of stop, λ is defined to be

$$\begin{pmatrix} \text{mean number of stops} \\ \text{per machine running} \\ \text{hour} \end{pmatrix} \times \begin{pmatrix} \text{mean service-} \\ \text{time per stop,} \\ \text{in hours} \end{pmatrix} = \mu\alpha, \text{ say.} \quad (1)$$

If there are several types of stop, λ is calculated by adding up (1) for all types of stop; in an obvious notation

$$\lambda = \mu_1 \alpha_1 + \mu_2 \alpha_2 + \dots \quad (2)$$

The quantity λ is the average number of hours of attention by the operative required to keep a single machine running for one hour. It is easily seen that an average running efficiency for one machine of $1/(1 + \lambda)$ would be obtained if no production was lost from interference.

There are two ways of measuring λ. The first is to estimate by time-study all the quantities in the expression (2). If estimates $\hat{\mu}_1, \hat{\alpha}_1, \ldots$, etc., are obtained subject to independent random sampling errors, the variance of the estimated servicing factor is given approximately by

$$\text{var}(\hat{\lambda}) \simeq \hat{\mu}_1^2 \text{var}(\hat{\alpha}_1) + \hat{\alpha}_1^2 \text{var}(\hat{\mu}_1) + \ldots \tag{3}$$

This formula can be used to estimate how extensive a time-study is desirable in any particular case; there is not space to go into this here, but it will often be found that the major contribution to the error in estimating λ arises from the determination of the rates of occurrence of randomly occurring stoppages.

The second method of measuring λ is by the snap-round method. Suppose that observation of a group of machines over a period shows that they are

running	R per cent of the time
stopped, receiving attention	A per cent of the time
stopped, awaiting attention	W per cent of the time.

Then it is not difficult to see that the servicing factor is A/R.

The attention-time is defined to include *directed movement time*, i.e. time in which the operative is walking with the clear purpose of going to a machine requiring attention. This introduces a small subjective element into the measurement of attention-times. Also the directed movement time will in general increase somewhat as n increases and an allowance for this may have to be included.

It is convenient to take as the unit of production the amount that would be produced by one machine running without stops for one hour. *The rate of production per machine* (sometimes called the machine efficiency or the machine availability) is then defined as the production of a single machine in a long time T hours, divided by T. This quantity is less than or equal to one and is often expressed as a percentage. Similarly, the *rate of production per operative* is the total rate of production of the whole set of machines and so equals n times the rate of production per machine. Finally, the *operative utilization* is defined to be equal to the proportion of time for which the operative is engaged in attending to stopped machines. Thus to say in a particular case that the operative utilization is 70 per cent means simply that for 70 per cent of the time the operative is attending to stopped machines and that for the remaining 30 per cent of the time all the machines are running.

First consider the result of applying the conservation argument of section 2.2. On the average each hour of running by one machine requires λ hours of attention by the operative, i.e. for each hour's attention by the operative, there are $1/\lambda$ hours of running time by a machine. Therefore

$$\begin{pmatrix} \text{rate of production} \\ \text{per operative} \end{pmatrix} = \begin{pmatrix} \text{operative} \\ \text{utilization} \end{pmatrix} \times \frac{1}{\lambda} \qquad (4)$$

and also

$$\begin{pmatrix} \text{rate of production} \\ \text{per machine} \end{pmatrix} = \begin{pmatrix} \text{operative} \\ \text{utilization} \end{pmatrix} \times \frac{1}{n\lambda}. \qquad (5)$$

These important equations are direct consequences of the definitions and do not require detailed analysis of the process. A certain amount of useful information can be inferred directly from (4) and (5). Thus if, say, $\lambda = 0{\cdot}1$, the maximum rate of production per operative is 10 units per hour; if, owing to other duties, the operative can spend only 90 per cent of her time attending to stopped machines, the maximum rate of production per operative is similarly 9 units per hour. These rates are attained only as limits when n is large; to calculate the production for particular values of n we need more detailed calculations.

4.2 The simplest probability model

We begin by examining the simplest case, in which

(*a*) on any one machine, stops occur completely randomly in running time, at the same rate α for all machines and independently for all machines;

(b) the frequency distribution of service-time is exponential, with p.d.f. $\sigma e^{-\sigma x}$, and therefore with mean $1/\sigma$. The service-times for different stops are statistically independent, and the service-times are independent of the number of machines awaiting attention;

(c) there is one server, always available for restarting stopped machines. That is, the system has unit capacity and complete availability.

In accordance with the discussion of section 1.3 (ix), (a) implies that if r machines are stopped at any time, the chance that a further machine stops in the next small time interval δt is $(n-r)\alpha\delta t + o(\delta t)$. Further, (b) implies that if at any time, a machine is being serviced, there is a constant chance $\sigma\delta t + o(\delta t)$, that the servicing operation is completed in the next δt, independently of how long the operation has been going on. Note that the servicing factor is $\lambda = \alpha/\sigma$.

Assumptions (a) and (b) are very severe; they will be discussed in detail in the next section.

The problem can be tackled by the method of section 2.4 and in fact the solution has been recorded there in subsection (vi). The formula for the equilibrium probability that r machines are stopped can be written in various ways. The simplest for numerical work is

$$p_0 = \frac{1}{F(\lambda, n)}, \qquad p_r = \frac{n!}{(n-r)!}\lambda^r p_0, \qquad (6)$$

where

$$F(\lambda, n) = 1 + n\lambda + n(n-1)\lambda^2 + \ldots + n!\,\lambda^n; \qquad (7)$$

the function $F(\lambda, n)$ is easily computed from the equations

$$F(\lambda, 1) = 1 + \lambda, \qquad F(\lambda, n+1) = (n+1)\lambda F(\lambda, n) + 1. \qquad (8)$$

Alternatively, we can write

$$p_r = \frac{e^{-1/\lambda}(1/\lambda)^{n-r}/(n-r)!}{\sum\limits_{u=0}^{n} e^{-1/\lambda}(1/\lambda)^{n-u}/(n-u)!}; \qquad (9)$$

this shows that the number of machines running, $n-r$, follows a truncated Poisson distribution of parameter λ^{-1}.

These results are due to Palm. From them we can calculate immediately the quantities of interest. For the operative utilization is the proportion of time that at least one machine is stopped and is therefore

$$1 - p_0 = 1 - \frac{1}{F(\lambda, n)}. \tag{10}$$

The rates of production per operative and per machine are then given by (4) and (5).

The most convenient way of tabulating the solution is by giving the operative utilization as a function of n and λ, and several such tables are available. Benson and Cox[*] gave a table up to $n = 15$ and Cox[†] gave a concise extension of this to large values of n. Table 4.1 is a short table which is adequate for approximate calculations with $n < 15$.

Example. Suppose that in a textile winding process a bobbin has to be replaced every 4·2 running-minutes and that the mean service-time is 16·3 seconds. In addition there is a yarn breakage on the average every 3·5 running-minutes and the mean service-time for this is 12·2 seconds. The servicing factor λ is, by (2),

$$\lambda = \frac{16 \cdot 3}{60} \times \frac{1}{4 \cdot 2} + \frac{12 \cdot 2}{60} \times \frac{1}{3 \cdot 5} = 0 \cdot 123.$$

We need to make a number of approximations before we can use Table 4.1. The first type of stop occurs regularly in running-time, whereas the second type can reasonably be assumed random. To apply the tables, we assume that this mixture of regular and random stops is reasonably approximated by a random series. This is likely to be so (see section 4.3 (iii)). Secondly, we assume both that the distribution of service-times when both types of stop are combined is nearly exponential, and also that no priority is given to one type of stop when selecting a machine on which to work.

[*] F. Benson and D. R. Cox, *J. R. Statist. Soc.*, B, **13** (1951), 65.
[†] D. R. Cox, *J. R. Statist. Soc.*, B, **16** (1954), 285.

TABLE 4.1

Operative utilization with n machines per operative and servicing factor λ.

n	λ 0·8	0·7	0·6	0·4	0·2
1	0·444	0·412	0·375	0·286	0·167
2	0·742	0·704	0·658	0·528	0·324
3	0·903	0·876	0·840	0·718	0·470
4	0·971	0·958	0·938	0·850	0·602
6	0·998	0·997	0·994	0·972	0·808
8	1·000	1·000	1·000	0·997	0·930
10	—	—	—	1·000	0·982
12	—	—	—	—	0·997

n	λ 0·18	0·16	0·14	0·12	0·10
2	0·298	0·271	0·242	0·212	0·180
4	0·560	0·515	0·465	0·412	0·353
6	0·767	0·718	0·660	0·592	0·516
8	0·902	0·864	0·813	0·746	0·662
10	0·969	0·949	0·916	0·863	0·785
12	0·993	0·986	0·970	0·939	0·880
14	0·999	0·997	0·992	0·978	0·943

Then, on making an approximate linear interpolation in Table 4.1, we obtain the operative utilizations given in Table 4.2. The rates of production per operative and per machine are then determined from (4) and (5).

Given the running speeds of the machines, the rates of production are easily converted into lb of yarn per hour. As is qualitatively obvious, production per operative increases, and production per machine decreases, as *n* is increased.

One practical use of the predictions in Table 4.2 would be in conjunction with cost data to decide on an optimum value of *n*. Another application is to decide whether a modification to the process, e.g. an increase in bobbin size, is likely to be justified. A new servicing factor is worked out, which is usually smaller than the old one, and the corresponding rates of production found. The increase in production due to the reduction in λ can be set against the cost of the modification; of course full advantage of such a modification can usually be obtained only by changing n. Before any such modification is put into effect on a large scale, it will be very desirable indeed to have a small-scale trial under as realistic conditions as possible, observing what happens in sufficient detail

TABLE 4.2

Production in a textile winding process.

n	Operative utilization	Rate of production per operative	Rate of production per machine
6	0·602	4·89	0·815
8	0·756	6·15	0·769
10	0·871	7·08	0·708
12	0·944	7·67	0·639

to determine not only by how much the prediction is in error, but also the reasons for the discrepancy.

A generalization of the solution can be obtained by supposing the system is of capacity m, i.e. by supposing there are m operatives any of whom may work on any machine in the set. The method of section 2.4 still holds, but now the probability that a servicing operation is completed in δt is $r\sigma\delta t + o(\delta t)$ if $r \leqslant m$ and is $m\sigma\delta t + o(\delta t)$ if $r > m$, where r is the number of machines stopped. It is left as an exercise for the reader to obtain and solve the appropriate equations, and to show that the analogue of (5) is that

$$\begin{pmatrix} \text{rate of production} \\ \text{per machine} \end{pmatrix} = \begin{pmatrix} \text{average operative} \\ \text{utilization} \end{pmatrix} \times \frac{m}{n\lambda}. \quad (11)$$

The average operative utilization is

$$\sum_{r=0}^{m} \frac{rp_r}{m} + \sum_{r=m+1}^{n} p_r, \qquad (12)$$

where p_r is the probability that r machines are stopped.

The following example gives some idea of the types of conclusion that can be drawn.

Example. Consider first a system with a fairly high servicing factor, say $\lambda = 0.45$. The first part of Table 4.3 gives the rate of

TABLE 4.3

Rates of production with a team of m operatives.

System	Operative utilization	Rate of production per operative	Rate of production per machine
(a) *High servicing factor*			
$\lambda = 0.45, n = 4, m = 1$	0·881	1·96	0·489
$n = 8, m = 2$	0·934	2·07	0·519
$n = 16, m = 4$	0·994	2·21	0·552
(b) *Low servicing factor, low loading*			
$\lambda = 0.05, n = 15, m = 1$	0·656	13·1	0·876
$n = 30, m = 2$	0·682	13·6	0·910
$n = 60, m = 4$	0·705	14·1	0·940
(c) *Low servicing factor, high loading*			
$\lambda = 0.05, n = 20, m = 1$	0·788	15·8	0·788
$n = 40, m = 2$	0·840	16·8	0·840
$n = 80, m = 4$	0·864	17·3	0·864

production per operative and per machine for various values of m and n. Parts (b) and (c) of the table show corresponding results for a system with a lower value for the servicing factor, λ.

For a given average number of machines per operative, i.e. for a given value of n/m, the rate of production per operative and per machine increases slowly as m increases. Of course there might well be practical considerations bearing on the choice of m as or more important than the theoretical changes in rate of production discussed here.

It can be shown that if both m and n are large with $n\lambda/m < 1$ the rate of production per machine approaches a limiting value $1/(1 + \lambda)$, whereas if $n\lambda/m > 1$, the asymptotic rate of production per machine is $m/(n\lambda)$.

4.3 Some complications

The solution discussed in the previous section is based on quite restrictive assumptions and there are numerous complications arising in practical applications which can make these assumptions seriously untrue. The mathematical study of what happens in such cases can soon get complicated, and we often have to be content with semi-empirical rules for modifying the solution of section 4.2. Benson* has made an intensive study of various complications that arise in practice and much of the following brief account is based on his work. Many of the points that arise in the following discussion occur in much the same way with other queueing problems.

(i) *Distribution of service-time*

It will frequently happen that the distribution of service-time is far from exponential. If only one type of operation is involved, the coefficient of variation of service-time will often be much less than 100 per cent, the value for the exponential distribution. When there are a number of types of stop, with a frequently occurring type with short service-time, and less frequently occurring stops with long service-times, the pooled distribution is likely to be nearer the exponential form, but the simple theory of section 4.2 is then applicable only if no important priorities exist.

A number of writers have given solutions for arbitrary distri-

* F. Benson, Unpublished Ph.D. thesis, University of Birmingham, 1957.

H

butions of service-time, when $m = 1$ and the other assumptions of section 4.2 are retained. Unfortunately the general solutions are not very convenient for numerical work, unless n is small, and the only tables available are those of Ashcroft*, which give, for $n < 20$, the rate of production per machine when the service-time is constant. For a fixed value of the servicing factor, the rate of production increases as the dispersion of service-time decreases. Analogy with Pollaczek's formula for the simple queue, equation (2.22), suggests that the operative utilization should be approximately a linear function of C_s^2, the square of the coefficient of variation of service-time, and that therefore linear interpolation on C_s^2 should be used between the values for $C_s^2 = 0$, constant service-time, and for $C_s^2 = 1$, exponentially distributed service-time. In many cases the difference between the predicted operative utilizations at $C_s^2 = 0, 1$ is not large, especially when comparative results only are required.

For example, in the situation of Table 4.2, the rates of production per operative with $n = 6, 8, 10, 12$ for constant service-time, $C_s = 0$ are, from Ashcroft's tables, 5·07, 6·44, 7·44 and 7·96. The corresponding values from Table 4.2, for exponentially distributed service-time, are 4·89, 6·15, 7·08, 7·67; the differences are small considering that a large change in the frequency distribution is involved.

(ii) *Availability of the operative*

In many practical applications the most unrealistic assumption in the analysis of section 4.2 is the one that the operative is always available to restart stopped machines, that is that the restriction on service is purely one of capacity and not one of availability. In fact, in machine interference problems, some allowance must always be made for relaxation and personal needs, and quite often also the operative will have other duties, such as fetching raw material. We call all demands on the operative's time, other than the restarting of stopped machines, *ancillary work*. A full discussion of the effect of ancillary work on the rate of production would

* H. Ashcroft. *J. R. Statist. Soc.*, B, **12** (1950), 145.

require a knowledge not only of the rate of incidence and duration of periods of ancillary work, but also of the priorities to be given to work of different types. It is nearly always impossible to specify these things at all precisely. Therefore the most sensible thing is to look for semi-empirical rules for modifying the solutions of section 4.2. The following is an outline of such an approach.

First suppose that the operative utilization calculated in the absence of ancillary work is U. Then the operative can spend up to a fraction $1 - U$ of his time on other duties, without loss of production, provided that the ancillary work is the sort that can be fitted into the periods during which all machines are running. Such ancillary work is called *tied*, because the times at which it is done are dictated by the machines. In general, if the proportion of time to be spent on such work is A_t, the proportion of time spent on productive work will be the smaller of $1 - A_t$ and the value read from the tables.

A second type of ancillary work whose effects are allowed for fairly easily is called *concentrated*. This is work done infrequently, but such that every time it is done the operative is unable to restart stopped machines for a considerable period. A rough guide is that the interval between successive items of such work is at least several hundred times the mean service-time for restarting stopped machines. A reasonable approximation in such cases is to suppose that no production is obtained during the periods of concentrated work, but that the normal rate of production applies in the remaining time. In fact some production will be obtained during a stretch of concentrated work, but this is roughly counterbalanced by the additional loss of production when work on the machines is begun again and the back-log of unattended machines is tackled.

In assessing the amount of concentrated work, we must distinguish between work which is proportional in amount to the total production and work which is independent of the total production. Thus an allowance for fetching material from a distant source would be of the first type, whereas an allowance for personal needs or for cleaning machines would be of the second. Thus a 5 per cent allowance for concentrated work reduces the production as estimated from Table 4.1 by 5 per cent. On the other hand, suppose that in the example of section 4.2 each unit* of production requires one minute of concentrated work. Then if A_c is the proportion of time spent on concentrated work, the rate of production per operative with, say $n = 10$, is, from Table 4.2, $7·08 (1 - A_c)$. Therefore this number of minutes of concentrated work is required per hour, so that

$$60A_c = 7·08(1 - A_c),$$

from which $A_c = 0·106$ and the rate of production per operative is 6·33. It is assumed that the work is done at infrequent intervals, e.g. once or twice a day.

A third type of ancillary work is called *spread* and can be thought of as done in a large number of small units of time spread uniformly over the whole period of operation. The effect of this is found approximately by defining a second servicing factor $\lambda' = \lambda/(1 - A_s)$, where λ is the servicing factor calculated as in section 4.1, ignoring ancillary work, and A_s is the proportion of time spent on spread ancillary work. In other words we imagine each piece of work at the machines to be extended by an appropriate factor. Table 4.1 is entered with servicing factor λ' and the value U' of the operative utilization gives the time spent on productive and spread ancillary work. The proportion of time spent on the former is approximately $U = (1 - A_s) U'$, and the rate of production is then derived from (1), using the original servicing factor λ.

It is, of course, possible to make a more refined classification of ancillary work, and indeed this would be advisable if the practical conclusions depended at all critically on the adjustment for ancillary work. There have been a number of interesting applications of the above ideas in the wool textile industries, which provide illustrations of the various kinds of ancillary work we have described. We give now an example, based closely on one given by Benson, Miller and Townsend.[†]

* I.e. the production of one machine running continuously for one hour.
† F. Benson, J. G. Miller, and M. W. H. Townsend, *J. Text. Inst.*, **44** (1953), T619.

Example. An operative has charge of a group of automatic pirn winders, transferring yarn from a large supply cone to small pirns. The duties of the operative are

(*a*) to replace the supply cone when empty;

(*b*) to repair yarn breaks that occur haphazardly from time to time;

(*c*) to replenish the magazines holding empty pirns.

The operative is allowed 5 per cent of her time for personal needs (considered as a concentrated ancillary allowance), and 10 per cent for supervision to be done while all machines are running (tied ancillary allowance). The service-times and frequencies of occurrence of tasks (*a*)–(*c*) are known.

The filling of magazines is done while the machines are running and therefore does not contribute to the primary servicing factor λ. One can consider three different methods of working, according to which the filling of magazines is treated as

(α) spread ancillary work;

(β) a mixture of spread and concentrated ancillary work;

(γ) tied ancillary work.

In (α) it is assumed that the operative attends frequently to the filling of magazines, putting one or two pirns into each magazine as the machines uses them. A revised servicing factor λ' is calculated and the number of machines is found to ensure that the operative spends 85 per cent of her time restarting stopped machines and filling magazines.

In (β) it is assumed that the operative waits until all magazines are nearly empty. She then fills all magazines, not attending to stopped machines while doing this. It turns out that the interval between successive items of such work is about 100 times the mean time to restart a stopped machine; in accordance with an empirical rule which has been suggested, such work is regarded as an equal mixture of spread and concentrated ancillary work. (Work is regarded as purely concentrated only if the interval is at least several hundred times the mean time to restart a stopped machine.)

In (γ) it is assumed that the filling of magazines is, so far as possible, fitted into those periods during which no machine is stopped.

Table 4.4 summarizes the numerical conclusions. The important general points are, first that the maximum rate of production per operative is fixed solely by the proportion of the operative's time free for productive work and is independent of the method of

TABLE 4.4

Production in pirn winding.

Method of working	Maximum rate of production per operative	No. of spindles per operative to achieve this	Rate of production per spindle	Rate of production per operative with 15 spindles
(α) Pirns inserted singly and often, whether machines are running or not	13·7	18	0·76	12·0
(β) Magazines all filled together when nearly run out	13·7	25	0·55	9·0
(γ) Pirns inserted singly when operative has no other duties	13·7	15	0·92	13·7

working. Secondly, (γ) is the best method of working in that the maximum is attained from the smallest number of machines and (β) is much the least efficient method. The ordering of methods (α), (β), (γ) is of course qualitatively obvious, but it is of value to express the differences quantitatively.

(iii) *Arrival pattern*

The theory of section 4.2 is based on a random occurrence of stops in the running-time of each machine. In the study of the simple queue (Chapters II and III) the form of the arrival pattern is quite important and in particular the congestion for regular arrivals is much less than for random arrivals. In the machine interference problem things are different. Even if stops occur regularly in the running-time of individual machines, the overall incidence of stops is the result of pooling several sequences of events, one sequence from each machine. It is known that unless the separate sequences keep in phase, the combined sequence will have many of the properties of a completely random series. Thus the solution of section 4.2 is likely to be a good approximation unless the operative can set up a regular cycle of working, which will be possible only when all or nearly all the stops occur regularly, when the service-time has small coefficient of variation and when n is small. In the extreme case when all stops occur regularly and the service-time is constant, the machines can be dealt with in a regular cycle and the operative utilization is

$$\text{Min}\left(1, \frac{n\lambda}{1+\lambda}\right), \tag{13}$$

which can be appreciably greater than the value from Table 4.1. It would not be difficult to modify (13) to account for the existence of a variability of service-time, small enough not to disturb the regular order of work at the machines; one can also fairly easily allow for occasional randomly occurring stops.

In the majority of applications, however, the variabilities of service-time and arrival pattern are likely to be sufficient to justify the use of results based on randomly occurring stops. There is scope for further research work, however, to determine more explicitly when this is justifiable.

(iv) *Absence of statistical uniformity*

The points to be discussed in the present subsection can be paralleled in many applications of queueing theory. For example, while the arrivals may be completely random over quite long time periods, the mean rate of arrival may vary in time. In some applications this may be systematic variation through the day; with machine stops it is more likely to be superimposed random variation between different batches of raw material, etc. Another source of non-uniformity is that the service-times for work of the same type may vary substantially between operatives, or, because of fatigue, may vary systematically in time.

In both cases there results a variation, between times or between operatives, in the servicing factor λ. What can be done about this depends on the particular application. If the service facilities have to be fixed and cannot be adjusted to follow variations in demand, it may be necessary to plan for the largest servicing factor that is likely to arise, providing additional work of another type during slack periods if possible. This approach may be desirable when customers are people, for then excessive congestion, even for localized periods, is not acceptable. (In the machine interference problem, the more serious difficulty, however, is often likely to arise from variations between operatives.)

If only the long-run production of the system is of interest, the use of an average servicing factor will lead to a slight overestimate. On the other hand, in the simple queue, the congestion may be seriously underestimated by using an average servicing factor.

(v) *Priorities*

It has been assumed that the selection of a particular machine for service is uncorrelated with the service-time that will be necessary to restart that machine. There are, however, various ways in which this assumption may fail.

First there may be two or more types of failure, which can be identified by the operative and which have different mean service-times. In a non-preemptive priority system the operative, when selecting a new machine on which to work, chooses the one with lowest mean service-time. In a preemptive priority system, the operative abandons work of low priority as soon as a machine stoppage of high priority occurs. As a general rule non-preemptive priority is unlikely to lead to appreciable increase in production but under some circumstances preemptive priority can; work with small mean service-time should be given priority over work with large mean service-time. There is unlikely to be worthwhile gain, and there may indeed be a loss of production, if the time taken to walk from one machine to another is comparable with the mean service-time.

The second possibility is that the machines may be divided into two or more groups, the groups have different servicing factors and possibly being such that the monetary value of a unit of production is different between groups. Provided that all occurrences of stops are random and that all service-times are exponentially distributed, we may study by the general methods of section 2.3 such questions as the relative advantages of different strategies.

A third and important form of priority may be based on the spatial arrangement of the machines. If an appreciable component of the service-time is the time to walk between machines, the servicing-factor is reduced by correct choice of the machine to be serviced whenever a number of machines are stopped; the simplest strategy is to work on the nearest machine.

(vi) *Patrolling*

In some applications, namely where the machines are spread over a considerable area and where it is important to examine each machine frequently whether or not it has stopped (spread ancillary work), the operative may adopt a scheme of regular patrolling. We can imagine that the machines are spaced round a circle and that the operative walks round the circle spending a time t_1 at machines that are running, and a time t_2 restarting any machine that is stopped. The essential difference from the situation studied earlier is that a stopped machine may remain unattended for a considerable time even though no other machines are stopped.

(vii) *Correlation of service-times with number of machines stopped*

It may happen that the operative tends to work quickly when the number of machines awaiting attention is large. In general, a tendency for the service-time to be shorter when the number of machines awaiting attention is large will increase the rate of production. However, such a tendency might well lead to a loss of quality.

CHAPTER V

More Specialized Topics

5.1 Introduction

In this chapter we deal briefly with a number of unrelated special topics, chosen partly to show some of the more complicated mathematical problems connected with queueing, and partly to illustrate a few problems of practical importance. The subjects dealt with are:

(a) the method of stages, a device for dealing in a simple way with problems involving a certain type of non-exponential distribution (section 5.2);

(b) a method of studying queueing problems by the use of integral equations (section 5.3);

(c) simulation and Monte Carlo methods (section 5.4);

(d) series of queues (section 5.5);

(e) the busy-period distribution (section 5.6).

5.2 The method of stages

We have seen repeatedly in earlier sections the great simplification introduced whenever a probability distribution connected with the queueing process is exponential. Thus if the distribution of service-time is exponential, with p.d.f. $\sigma e^{-\sigma x}$, there is a constant probability $\sigma \delta t + o(\delta t)$ that a service operation in progress at t is completed in $(t, t + \delta t)$, quite independently of how long the operation has been in progress. If all the random variables determining the system are independently exponentially distributed, the method of section 2.2 can be used to obtain simultaneous linear equations for the equilibrium probabilities, and simultaneous linear differential equations with constant coefficients for the non-equilibrium distribution.

In his pioneer work on congestion in telephone systems, A. K. Erlang suggested an ingenious method by which the simple arguments used for exponential distributions can be applied also for certain non-exponential distributions. The general idea has been outlined in section 1.4 (i) (c). In the simplest form of the method, applied to the distribution of service-time, we suppose there to be k stages of service, each with a service-time having the exponential p.d.f. $\sigma e^{-\sigma x}$. The stages need have no physical meaning. When stage 1 of service is complete the customer passes immediately to stage 2, and so on, the whole servicing operation being complete when the kth stage of service is complete. The total service-time is thus the sum of k independent identically distributed random variables with the p.d.f. $\sigma e^{-\sigma x}$. The Laplace transform of the exponential distribution is $\sigma/(\sigma + s)$, so that the Laplace transform for the total service-time is $\sigma^k/(\sigma + s)^k$. If we write b_1 for the mean service-time, then $\sigma = k/b_1$, and we may write for the p.d.f. of service-time

$$\frac{(k/b_1)\,(kx/b_1)^{k-1}e^{-kx/b_1}}{(k-1)!}. \tag{1}$$

The coefficient of variation of service-time is $1/\sqrt{k}$; in the limiting case $k \to \infty$ we obtain constant service-time.

To illustrate the method of stages we shall apply it to a simple illustrative example. Consider a single-server queue with random arrivals at rate α and with service-time distribution of the special Erlangian form (1). Properties of this system can be obtained for general distributions of service-time from the results of section 2.6, so that the use of the method of stages is not in fact necessary here. If the distribution of service-time were exponential we would define the state of the system by the number of customers in the system. To deal with the more general distribution we must extend the definition of the state of the system to include a specification of the stage of service in progress.

The most direct way to do this is to denote by $p_{ni}(t)$, the probability that at time t there are n customers in the system and that the customer at the service point is in the ith stage of service

$(n = 1, 2, \ldots; i = 1, 2, \ldots, k)$. Let $p_0(t)$ be the probability that there are no customers in the system at time t. Then by considering the transitions that can occur in a small time interval $(t, t + \delta t)$, we have equations of which the simplest is

$$p_0(t + \delta t) = p_0(t)(1 - \alpha \delta t) + \sigma p_{1k}(t) \delta t.$$

Note that on the right-hand side σ occurs in combination with p_{1k} only; this is because the state 0 can be entered only when the final, k^{th}, stage of service of a single customer is completed. If we consider equilibrium probabilities the full set of equations is

$$\left.\begin{aligned}
\alpha p_0 &= \sigma p_{1k}, \\
(\alpha + \sigma) p_{11} &= \alpha p_0 + \sigma p_{2k}, \\
(\alpha + \sigma) p_{1i} &= \sigma p_{1, i-1} \quad (i = 2, \ldots, k), \\
(\alpha + \sigma) p_{n1} &= \alpha p_{n-1, 1} + \sigma p_{n+1, k} \quad (n = 2, \ldots), \\
(\alpha + \sigma) p_{ni} &= \alpha p_{n-1, i} + \sigma p_{n, i-1} \quad (n = 2, \ldots; i = 2, \ldots, k).
\end{aligned}\right\} \quad (2)$$

The important general point, applying to all applications of the stage device, is that, just as when the distribution of service-time is exponential, the equilibrium distribution is obtained by solving simultaneous linear equations, which can be written down in a routine way. Even if it is difficult to solve the equations explicitly, numerical solution on a computer will usually be possible.

In order to solve the equations (2) it is helpful to change the notation by writing $p_{ni} = p_{nk-i+1}$; that is, the two-component suffix (n, i) is replaced by a single-component suffix. The interpretation of this transformation is that $nk - i + 1$ is the number of stages awaiting service, including the one in course of service. The new equilibrium equations are

$$\left.\begin{aligned}
\alpha p_0 &= \sigma p_1, \\
(\alpha + \sigma) p_r &= \sigma p_{r+1} + \alpha p_{r-k} \quad (r = 1, 2, \ldots),
\end{aligned}\right\} \quad (3)$$

with the understanding that any p with negative suffix is zero. Equations (3) can be reinterpreted as the equations of a simple queue with bulk arrivals. For if groups of customers, each containing k customers, arrive at random at rate α, and if the service-time of single customers is exponential with parameter σ, then equations (3) determine the equilibrium probability p_r of there being r customers in the system at an arbitrary time-point. It is clear, in fact, that in the new system customers correspond to stages in the original model.

Equations (3) can be combined into a single equation by defining a generating function

$$G(\zeta) = \sum \zeta^r p_r.$$

Then, when we multiply the equation (3) for p_r by ζ^r and add over r, we have that

$$G(\zeta) = \frac{\sigma p_0(1-\zeta)}{\sigma + \alpha\zeta^{k+1} - (\alpha+\sigma)\zeta}. \tag{4}$$

To determine p_0 we may use the normalizing equation

$$G(1) = \sum p_r = 1.$$

On evaluating the right-hand side of (4) for $\zeta = 1$ by the method of indeterminate forms, we have that

$$\begin{aligned} p_0 &= 1 - \alpha k/\sigma, \\ &= 1 - \alpha b_1, \\ &= 1 - \rho, \end{aligned}$$

where $\rho = \alpha b_1$ is the traffic intensity. An alternative proof is obtained from a consideration of the long-run proportion of time for which the server is occupied. Thus

$$G(\zeta) = \frac{(1-\rho)(1-\zeta)}{\left(1 + \frac{\rho}{k}\zeta^{k+1}\right) - \left(1 + \frac{\rho}{k}\right)\zeta}. \tag{5}$$

If we want the probability that there are n customers in the system we must calculate

$$\sum_{r=(n-1)k+1}^{nk} p_r \quad (n = 1, 2, \ldots). \tag{6}$$

Note that this gives the long-run proportion of time for which there are n customers in the system, whereas the probability distribution obtained in section 2.6 gave the long-run proportion of those time instants at which service is just completed, for which n customers remain in the system.

The queueing-time of a customer can be found from the consideration that with probability p_r the queueing-time consists of the sum of r exponentially distributed quantities. This leads to a special case of the result of section 2.6, equation (28).

The same type of argument can be used when the intervals between successive arrivals are independently distributed with a special Erlangian distribution formed from k_a stages. We then suppose that on the arrival of a customer the arrival 'mechanism' enters the first stage. At the end of the first stage, the 'mechanism' passes to the second stage, and so on, until at the end of the $(k_a)^{th}$ stage the next customer arrives. If the times spent in the separate stages are independently exponentially distributed, the intervals between successive arrivals are independently distributed in the required special Erlangian distribution.

To formulate the equations of the queueing process the specification of the state of the system must include a statement of the stage occupied by the arrival 'mechanism'. Thus to deal by this method with a single-server queue with exponential service-time and with independent special Erlangian arrivals, we denote by p_{ni} ($n = 0, 1, \ldots$; $i = 1, \ldots, k_a$) the equilibrium probability that there are n customers in the system and that the arrival 'mechanism' is in the ith stage. Linear equations for the p_{ni} are then formed in the usual way. To take a more complicated example, suppose that there are m servers, that the distribution of service-time is special Erlangian with k_s stages, and that the distribution of intervals between successive arrivals is special Erlangian with k_a stages. We then define equilibrium probabilities as set out in Table 5.1.

Equilibrium equations can again be written down, although they will be of complicated form unless m and k_s are both small. One virtue of the method of stages is that it enables such equations to be written down in a routine way. If facilities are available for solving large sets of simultaneous linear equations, a numerical solution is thus obtained even for quite complicated systems. Before numerical solution is possible the set of equations in infinitely many unknowns has to be reduced to a finite set. For this we choose r_0 so that the probability of r_0 or more customers in the system is thought negligible and then set all the corresponding p's equal to zero. If practicable, the equations should be solved for several values of r_0 and the solution extrapolated to infinite r_0.

TABLE 5.1

Probabilities connected with m-server queue treated by method of stages.

Probability	Number of customers in system	Stage of arrival mechanism	Stage of service	Ranges of suffices
p_{0i}	0	i	—	$1 \leqslant i \leqslant k_a$
p_{1ij_1}	1	i	j_1	$1 \leqslant i \leqslant k_a; 1 \leqslant j_1 \leqslant k_s$
$p_{2ij_1 j_2}$	2	i	j_1, j_2	$1 \leqslant i \leqslant k_a;$ $1 \leqslant j_1 \leqslant j_2 \leqslant k_s$
$p_{rij_1 \cdots j_r}$	r	i	$j_1, \ldots, j_r$	$1 \leqslant i \leqslant k_a;$ $1 \leqslant j_1 \leqslant \ldots \leqslant j_r \leqslant k_s;$ $1 \leqslant r \leqslant m;$
$p_{rij_1 \cdots j_m}$	r	i	$j_1, \ldots, j_m$	$1 \leqslant i \leqslant k_a;$ $1 \leqslant j_1 \leqslant \ldots \leqslant k_s;$ $m \leqslant r$

Returning to the distribution (1), we note that it has coefficient of variation $1/\sqrt{k}$, and therefore represents for $k = 2, 3, \ldots$, a family of distributions with coefficient of variation less than that of the exponential distribution, $k = 1$. An immediate generalization of (1) is obtained by allowing the exponential distributions in the k stages to have parameters $\sigma_1, \ldots, \sigma_k$, not necessarily equal. Then the Laplace transform of total service-time is

$$\prod_{i=1}^{k} \frac{\sigma_i}{\sigma_i + s} \tag{7}$$

The corresponding coefficient of variation is

$$\frac{(\sum \sigma_i^{-2})^{1/2}}{(\sum \sigma_i^{-1})}$$

and is easily seen to lie between $1/\sqrt{k}$ and 1. Therefore no extension in the range for the coefficient of variation has been obtained by this generalization.

A different generalization arises if we suppose that with probability λ_i, $\sum \lambda_i = 1$, a customer has service-time that is exponentially distributed with a mean of $1/\sigma_i$. The Laplace transform of service-time is then

$$\sum_{i=1}^{k} \frac{\lambda_i \sigma_i}{\sigma_i + s}. \tag{8}$$

With $k = 2$, equation (8) can be written

$$\frac{\lambda \sigma_1}{\sigma_1 + s} + \frac{(1-\lambda) \sigma_2}{\sigma_2 + s}$$

and the coefficient of variation can be made arbitrarily large by choice of the adjustable parameters. Thus the device of stages in parallel rather than in series will be useful when it is required to represent distributions with a very high coefficient of variation. The specification of the state of the system must include the stage occupied by the customer being served.

From a mathematical point of view it is interesting that (7) is the reciprocal of a polynomial of degree k, whereas (8) is the ratio of two polynomials, the denominator having degree k. It can be shown that any distribution of a positive random variable, having a Laplace transform the ratio of two polynomials, can be represented by a combination of stages in series and in parallel, the number of stages being equal to the degree of the denominator. We call this the family of general Erlangian distributions.

If we wish to approximate to an empirical frequency distribution by an Erlangian distribution a reasonable procedure is to try first a distribution like (1), if the coefficient of variation is less than 1, and (8) if the coefficient of variation exceeds 1. If the fit is judged not good enough, we can try two distributions (1) in parallel, and so on until a good fit is obtained. A trial-and-error procedure of this sort is likely to be better than a purely formal procedure such as fitting by moments. It will normally be required to keep k as small as possible; k cannot be appreciably less than the square of the reciprocal of the coefficient of variation, if the first and second moments of fitted distribution are to be nearly equal to those of the empirical distribution.

5.3 The integral equation of a single-server queue

We have seen in earlier chapters that the stationary distribution of single-server queue-size can be obtained, without too much difficulty. However, our assumptions have always involved the presence of randomness. For we have either supposed that the customers arrive at random, or that the service-times have an exponential distribution, which amounts to the assumption that the points in time when service is completed are randomly distributed, at least while the server is busy.

In many single-server queueing situations our convenient assumptions fail to hold. It may be that service-times have a quite general distribution $B(x)$, and that the intervals between arrivals of customers, while independent, have a general distribution $A(x)$. A particular example covered by this more general set-up arises when the intervals between arrivals are of constant length, i.e. when items arrive regularly at the service-point.

We call the general arrangement involving the distributions $A(x)$ and $B(x)$ the single-server queue with general independent arrivals and general service-time. As we have mentioned, the methods developed in earlier chapters are not adequate for discussing this case. In the present section, therefore, we shall give a brief account of this new and more difficult problem. The solution involves a certain, somewhat intractable, integral equation, and we shall see how this equation can be solved in certain cases.

Let us write C_n for the nth customer, X_n for his queueing-time, Y_n for his service-time, and Z_n for the interval between the arrival of C_n and the arrival of C_{n+1}. Then X_n, Y_n, Z_n are independent quantities, as a moment's reflection will show. Indeed, Y_n and Z_n are independent of the entire history of the queueing-process up to the arrival of C_n. The random variable Y_n has the distribution $B(x)$, and Z_n has the distribution $A(x)$. We shall put $b_1 = EY_n$ and $a_1 = EZ_n$; thus ρ, the traffic intensity, is equal to b_1/a_1. We shall write $F_n(x)$ for the distribution of X_n. If $\rho < 1$ it is very reasonable to expect there to be a stationary distribution of queueing-time. The existence of the stationary distribution can be proved, but here we shall assume its existence and concentrate on the problem of its calculation. That is, we assume that as $n \to \infty$, $F_n(x)$ tends to a limit, $F(x)$, and consider the problem of calculating $F(x)$.

It is not difficult to see that X_n and X_{n+1} are related by the equation

$$X_{n+1} = X_n + Y_n - Z_n,$$

provided that the quantity on the right is positive. If $X_n + Y_n - Z_n$ is negative, the customer C_{n+1} arrives to find the server free, and so then we have, simply, $X_{n+1} = 0$. In general

$$X_{n+1} = \text{Max}(X_n + Y_n - Z_n, 0).$$

The following type of argument is one that can be attempted for any system described by a recurrence relation of this general nature. As a matter of fact, the numerical value of $X_n + Y_n - Z_n$, when it is algebraically negative, measures the time during which the server was free between the serving of C_n and the arrival of C_{n+1}. In either event, whether $X_n + Y_n - Z_n$ is positive or negative, it should be clear that provided that $x \geqslant 0$ we have

$$\text{prob}\{X_{n+1} \leqslant x\} = \text{prob}\{X_n + Y_n - Z_n \leqslant x\}. \tag{9}$$

To proceed further, let us put

$$K(x) = \text{prob}\{Y_n - Z_n < x\}$$

for the distribution function of the quantity $Y_n - Z_n$. Evidently

$$K(x) = \int\limits_0^\infty B(x+z)\, dA(z),$$

so that $K(x)$ is in principle calculable from known distributions. Further, since X_n and $Y_n - Z_n$ are independent, the distribution function of $X_n + Y_n - Z_n$ is

$$\int_0^\infty K(x-y)\,dF_n(y).$$

Thus (9) is telling us that for all $x \geqslant 0$ we must have

$$F_{n+1}(x) = \int_0^\infty K(x-y)\,dF_n(y).$$

If we let $n \to \infty$ in this equation we have that $F(x)$, the stationary queueing-time distribution, satisfies the equation

$$F(x) = \int_0^\infty K(x-y)\,dF(y), \tag{10}$$

for all $x > 0$. This equation may be called the integral equation of the queue. It is of the *Wiener–Hopf* type and its general solution requires an elaborate mathematical discussion. We shall content ourselves with seeing how (10) may be solved for certain interesting special cases.

To begin with, we shall show how solutions to (10) are to be obtained when the inter-arrival intervals can be regarded as the sum of several exponential variables. In other words, we shall suppose that if Z is a typical interval between arrivals, then we can put

$$Z = U_1 + U_2 + \ldots + U_k, \tag{11}$$

where $U_1, U_2, \ldots, U_k$ are independent exponential variables with means $\lambda_1^{-1}, \lambda_2^{-1}, \ldots, \lambda_k^{-1}$ respectively. The Laplace–Stieltjes transform of $A(x)$ is

$$A^\star(s) = \frac{\lambda_1 \lambda_2 \ldots \lambda_k}{(\lambda_1 + s)(\lambda_2 + s)\ldots(\lambda_k + s)}, \tag{12}$$

because it is just the product of k transforms of exponential distributions. Distributions like this arise also in Erlang's 'stage' device (section 5.2), and by choice of k and the λ's we can represent a wide range of inter-arrival distributions. Thus, although we shall not be presenting the most general solution to the queue-equation, the solution we do present should serve most practical purposes.

To understand our argument it is necessary to establish first a preliminary result. Suppose that U_1 is an exponential variable with mean λ_1^{-1}, and suppose that W_1 is a continuous variable independent of U_1. Let $p_1(x)$ be the density function of W_1 and let $q(x)$ be the density function of $W_1 - U_1$. Then

$$q(x) = \int\limits_x^\infty \lambda_1 e^{-\lambda_1(z-x)} p_1(z)\,dz,$$

from which it follows by straightforward differentiation that

$$\frac{d}{dx}q(x) = \lambda_1 q(x) - \lambda_1 p_1(x),$$

or
$$\left(1 - \frac{1}{\lambda_1}\frac{d}{dx}\right)q(x) = p_1(x). \tag{13}$$

In other words, the operator

$$1 - \frac{1}{\lambda_1}\frac{d}{dx}$$

applied to $q(x)$ removes its exponential component and gets us back to $p_1(x)$. Obviously, if we know that W_1 is representable as $W_2 - U_2$, where W_2 and U_2 are independent, and that U_2 is exponential with mean λ_2^{-1}, we can apply a similar operation to (13) and get

$$\left(1 - \frac{1}{\lambda_2}\frac{d}{dx}\right)\left(1 - \frac{1}{\lambda_1}\frac{d}{dx}\right)q(x) = p_2(x),$$

where $p_2(x)$ is the density function of W_2.

Now the right-hand side of the fundamental equation (10) is the distribution function of the variable $X + Y - Z$, where X is a variable with the stationary queueing-time distribution, Y has the service-time distribution and Z has the inter-arrival distribution. Moreover X, Y, Z are independent. In view of our assumption (11) about Z, and of the results we have just obtained about our operators, it should be evident that if we apply the operator

$$\left(1 - \frac{1}{\lambda_1}\frac{d}{dx}\right)\left(1 - \frac{1}{\lambda_2}\frac{d}{dx}\right)\cdots\left(1 - \frac{1}{\lambda_k}\frac{d}{dx}\right)$$

to the density function of $X + Y - Z$ we must get the density function of $X + Y$. If we suppose Y to be a continuous variable with density $b(x)$, then the density function of $X + Y$ is

$$\int_0^x b(x - z)\,dF(z).$$

But the queue-equation is telling us that for all $x > 0$ the density function of X is the same as the density function of $X + Y - Z$. Thus we obtain the result that for all $x > 0$

$$\left(1 - \frac{1}{\lambda_1}\frac{d}{dx}\right)\left(1 - \frac{1}{\lambda_2}\frac{d}{dx}\right)\cdots\left(1 - \frac{1}{\lambda_k}\frac{d}{dx}\right)f(x) = \int_0^x b(x - z)\,dF(z). \quad (14)$$

The reader will notice that we have written $f(x)$ for the density function of X, for $x > 0$, but have written the right-hand side of (14) in the form of a Stieltjes convolution. The reason is as follows. The queueing-time has a positive probability of being zero, in general. Thus, $F(x)$ has a definite 'jump' at the origin, the saltus there measuring the probability of a customer not being obliged to wait for service. Under the assumption that Y and Z are continuous, however, the distribution of queueing time *conditional upon its being strictly positive* is continuous. In other words, we can put $f(x) = F'(x)$ for $x > 0$, but $F'(x)$ does not exist at $x = 0$.

If we let P be the probability of zero queueing-time, then (14) can be rewritten

$$\left(1 - \frac{1}{\lambda_1}\frac{d}{dx}\right)\left(1 - \frac{1}{\lambda_2}\frac{d}{dx}\right)\cdots\left(1 - \frac{1}{\lambda_k}\frac{d}{dx}\right)f(x)$$

$$= Pb(x) + \int_0^x b(x-z)f(z)dz. \tag{15}$$

At this stage it may be wondered why so much trouble has been taken to convert a comparatively simple-looking equation like (10) into the considerably more complicated-looking version (15). The reason is that the latter version (15) is a form which is perfectly suited to the application of the familiar one-sided Laplace transform. It is to be recalled that the Laplace transform of $f^{(r)}(x)$ is

$$s^r f^0(s) - s^{r-1}f(0+) - s^{r-2}f'(0+) - \ldots - f^{(r-1)}(0+),$$

where the symbol $0+$ means that limiting values are to be taken as the argument decreases to zero through positive values, and where

$$f^0(s) = \int_0^\infty e^{-sx}f(x)\,dx.$$

If we take Laplace transforms of (15), we thus obtain

$$\left(1 - \frac{s}{\lambda_1}\right)\left(1 - \frac{s}{\lambda_2}\right)\cdots\left(1 - \frac{s}{\lambda_k}\right)f^0(s)$$

$$= Pb^0(s) + b^0(s)f^0(s) + P_{k-1}(s), \tag{16}$$

where $P_{k-1}(s)$ is a polynomial in s, of degree $k-1$, the coefficients of which involve the unknown quantities $f(0+)$, $f'(0+)$, ..., $f^{(k-1)}(0+)$. After a little algebra (16) simplifies to

$$P + f^0(s) = \frac{Q_k(s)}{[a^0(-s)]^{-1} - b^0(s)}, \tag{17}$$

where $Q_k(s)$ is an unknown polynomial in s, of degree k.

To determine $Q_k(s)$ it is necessary to make use of a little theory of complex variables. If we suppose s to be a complex number, $f^0(s)$ must be finite whenever $\mathscr{R}s \geqslant 0$. This is because $f^0(0)$, the total integral of $f(x)$, is finite, and because, when $\mathscr{R}s \geqslant 0$, the modulus of the integrand defining $f^0(s)$ is less than or equal to that for $f^0(0)$. Now it is possible to show that, when $\rho < 1$, there are exactly $k-1$ zeros of $[a^0(-s)]^{-1} - b^0(s)$ whose real part is strictly positive. In other words, there are exactly $k-1$ complex numbers $s_1, s_2, ..., s_{k-1}$ such that

$$a^0(-s_j)b^0(s_j) = 1 \quad (j = 1, 2, ..., k-1),$$
and
$$\mathscr{R}s_j > 0 \quad (j = 1, 2, ..., k-1).$$

Furthermore $[a^0(-s)]^{-1} - b^0(s)$ obviously vanishes at $s = 0$.

Thus, in order for $f^0(s)$ to be finite for $\mathscr{R}s \geqslant 0$, $Q_k(s)$ must vanish at $0, s_1, s_2, ..., s_{k-1}$. Hence

$$Q_k(s) = Cs(s-s_1)(s-s_2)...(s-s_{k-1}),$$

where C is some constant. The constant C can be determined by letting s tend to zero in (17), for we must have $P + f^0(0) = 1$. To determine the limit of the right-hand side of (17) we need to use the rule of l'Hospital, and we find after a little calculation that $C = (b_1 - a_1) / \prod_{j=1}^{k-1} (-s_j)$. Thus

$$P + f^0(s) = \frac{(b_1 - a_1) s \prod_{j=1}^{k-1} \left(1 - \dfrac{s}{s_j}\right)}{[a^0(-s)]^{-1} - b^0(s)}. \tag{18}$$

This equation gives the Laplace transform of the stationary distribution of queueing-time. It involves no unknown quantities except for the $k-1$ zeros, which may of course be difficult to obtain in any particular situation. Notice that when $k = 1$, so that arrivals are random, we have no zeros to determine, and (18) then agrees with results we already have obtained for random arrivals (section 2.6).

The probability P that a customer does not have to queue can be determined by letting $s \to +\infty$ (through real values) in (18), because $f^0(s) \to 0$ as $s \to +\infty$. We find in this way that

$$P = (a_1 - b_1) \prod_{j=1}^{k} \lambda_j \bigg/ \prod_{j=1}^{k-1} s_j.$$

Consider next the following example

$$A^\star(s) = (1+s)^{-2}, \qquad B^\star(s) = e^{-s}.$$

This corresponds to a constant service-time of one unit and an inter-arrival density function

$$a(x) = e^{-x} x,$$

with mean two units. Thus $\rho = \frac{1}{2}$ and there should be a stationary queueing-time distribution. Clearly $A^\star(s)$ is the reciprocal of a quadratic function, so that the preceding theory holds with $k = 2$. We have to find the one zero of $(1-s)^2 - e^{-s}$ whose real part is strictly positive. This is $s_1 \simeq 1 \cdot 478$. Now $a_1 = 2, b_1 = 1, \lambda_1 = \lambda_2 = 1$. Thus $P \simeq (1 \cdot 478)^{-1} \simeq 0 \cdot 6766$, i.e. the probability is about 68 per cent that a customer will not have to queue.

This may be compared with the result (section 2.6; equation (19)) that with random arrivals the probability of zero queueing-time is $1 - \rho$; with $\rho = \frac{1}{2}$, this gives a probability of 50 per cent. As might be expected the probability of not having to queue is greater in the more regular system.

To complete our example we can show that

$$0 \cdot 6766 + f^0(s) \simeq \frac{s \left(\dfrac{s}{1 \cdot 478} - 1 \right)}{(1-s)^2 - e^{-s}}$$

This transform is not simple to invert, but expansion in powers of s leads easily to the moments of $f(x)$.

Let us now return to the general problem, and suppose that $b^0(s) = [\beta_l(s)]^{-1}$, where $\beta_l(s)$ is a polynomial of degree l. For the type of inter-arrival distribution we are considering we can write

$a^0(s) = [\alpha_k(s)]^{-1}$, where $\alpha_k(s)$ is a polynomial of degree k in s. We then have from (17) that

$$P+f^0(s) = \frac{Q_k(s)\beta_l(s)}{\beta_l(s)\alpha_k(-s)-1}. \qquad (19)$$

However, the denominator on the right-hand side is a polynomial of degree $k+l$ and so has $k+l$ zeros. We know that one of these zeros is at $s = 0$, and that there are exactly $k-1$ further zeros $s_1, s_2, \ldots, s_{k-1}$ whose real parts are positive. Thus the remaining l zeros, $z_1, z_2, \ldots, z_l$, say, must have negative real parts, and we must be able to write

$$\beta_l(s)\alpha_k(-s)-1 = D(s-z_1)(s-z_2)\ldots(s-z_l)Q_k(s),$$

where D is a constant. This last equation gives, from (19), that

$$P+f^0(s) = \frac{\beta_l(s)}{D\prod\limits_{j=1}^{l}(s-z_j)}.$$

The constant D is easy to find, like C, by letting $s \to 0$. We obtain

$$D^{-1} = \prod\limits_{j=1}^{l}(-z_j)$$

and hence have the simple result that

$$P+f^0(s) = \frac{\beta_l(s)}{\prod\limits_{j=1}^{l}\left(1-\frac{s}{z_j}\right)}. \qquad (20)$$

Since $\beta_l(s)$ is a polynomial of degree l, and since $[\beta_l(s)]^{-1} = b^0(s)$ must be finite when $\mathscr{R}s \geqslant 0$, we must have

$$\beta_l(s) = \prod\limits_{j=1}^{l}\left(1-\frac{s}{\zeta_j}\right),$$

where $\mathscr{R}\zeta_j < 0$ for $j = 1, 2, \ldots, l$. Thus, letting $s \to \infty$ in (20), we have that

$$P = \prod_{j=1}^{l} (z_j/\zeta_j). \tag{21}$$

It appears, therefore, that when $b^0(s)$ is the reciprocal of a polynomial of degree l in s we need only discover the l zeros of $a^0(-s)b^0(s) - 1$ whose real parts are negative. It is then easy to calculate P and $f^0(s)$. This procedure in no way depends upon k, the degree of $[a^0(s)]^{-1}$, which can presumably become arbitrarily large. Indeed, it seems on the basis of the results obtained, highly plausible that the procedure is valid for all $a^0(s)$. It has been proved that this plausible idea is actually correct, but we do not give the proof here. Thus, if $b^0(s) = [\beta_l(s)]^{-1}$, where $\beta_l(s)$ is a polynomial of degree 1 with zeros $\zeta_1, \zeta_2, \ldots, \zeta_l$, there are exactly l zeros $z_1, z_2, \ldots, z_l$ of $[a^0(-s)]b^0(s) - 1$ with negative real parts, P is given by (21), $f^0(s)$ by (20).

Consider the following example:

$$b^0(s) = \frac{1}{\left(1 + \dfrac{s}{3}\right)^2}, \qquad A^\star(s) = e^{-s}.$$

This means that arrivals are regular, one per unit time, while

$$b(x) = 9x e^{-3x} \quad (x \geqslant 0).$$

We require the two zeros z_1, z_2 of

$$\left(1 + \frac{s}{3}\right)^2 e^{-s} - 1$$

whose real parts are negative. Numerical computations show that

$$z_1 \simeq -1{\cdot}748, \qquad z_2 \simeq -3{\cdot}517.$$

Note that $\zeta_1 = \zeta_2 = -3$. Thus

$$P \simeq 0{\cdot}6831,$$

i.e. again there is a value of about 68 per cent for the probability of a customer not queueing. Since $b_1 = \frac{2}{3}$, $a_1 = 1$, the present example has $\rho = \frac{2}{3}$, so that $1 - \rho$ is much further from P than it was in the last numerical example.

One can show further that

$$P + f^0(s) \simeq \frac{0 \cdot 346}{\left(1 + \dfrac{s}{1 \cdot 748}\right)} - \frac{0 \cdot 0293}{\left(1 + \dfrac{s}{3 \cdot 517}\right)}$$

and thus

$$f(x) \simeq 0 \cdot 605\, e^{-1 \cdot 748x} - 0 \cdot 103\, e^{-3 \cdot 517x}.$$

This numerical example, with regular arrivals and a fairly simple distribution of service-time, could have been obtained by modifying the analysis of section 3.2 for regular arrivals and exponential service-time by the use of the stage device of section 5.2; service would be regarded as composed of two stages. A rather similar modification of the analysis for random arrivals and constant service-time would deal with the earlier numerical example. The advantage of the stage-device is that the argument, although messy, is elementary, and is applicable to a wide range of problems. The argument of the present section, on the other hand, leads both to elegant general results and to a clear statement of what needs to be calculated in any particular application. The present method is only applicable to systems governed by an integral equation of the same general type as (10), but where the method can be used, we prefer it.

5.4 Simulation and Monte Carlo methods

Very often, instead of making a mathematical analysis of the properties of the queueing system under study, it may be advisable to examine the process by reconstructing its behaviour using service-times, arrival times, etc., derived from random numbers. This approach is particularly useful when the process is so complicated that mathematical solution is likely to be difficult or impossible, and especially when the behaviour is required under very special and clearly defined conditions and no mathematical solution is immediately available. It may then happen that empirical sampling is likely to lead to an answer in a reasonable time, whereas the effort needed to produce a mathematical solution may be difficult to gauge.

The simplest procedure is to use the random numbers to construct a direct realization of the queueing process corresponding as closely as possible to the real system. In order to obtain more precise conclusions for a given amount of effort, it may, however, be profitable to modify the process that is sampled. There is considerable scope for subtlety in this. There is no universally accepted terminology, but the term *Monte Carlo method* is often reserved for a procedure in which the process sampled had been modified to increase precision, whereas the term *simulation* is used when the process sampled is a close model of the real system. An advantage of simulation over a Monte Carlo method is that the detailed results give a direct qualitative impression of what the behaviour of the system should look like under the conditions postulated.

The following example is intended to illustrate the steps in setting up a simulation study and some of the difficulties that can arise. The object of the example is to illustrate the method. Considerably more calculation would be necessary to reach an accurate answer, and, indeed, the reader who intends to use straightforward simulation must usually be prepared for some extensive computing. Of course, this may be unimportant if the calculation can conveniently be programed for an electronic computer.

Suppose that there are two servers, that the distribution of service-time is exponential with mean $1\frac{1}{2}$ units, that the queue-discipline is 'first-come, first-served' and that the arrival pattern is as follows. Arrivals are scheduled regularly at times 1, 2, ..., but the actual arrival points are independently normally distributed around the scheduled arrival points as mean and with standard deviation $\frac{1}{2}$ unit. The traffic intensity is $\frac{3}{4}$. Let it be required to estimate the mean queueing-time per customer.

The first step is to generate appropriately distributed random quantities. To produce observations independently distributed with distribution function $F(x)$, the procedure is as follows.

Let $y_1, y_2, \ldots$ be independently uniformly distributed over $(0, 1)$. For example they may be sets of say three digits read from a table of random digits*, each set interpreted as a three-decimal-place number in the range 0·000–0·999. Then let x_1 be the largest value of x for which $y_1 \geqslant F(x)$, etc. Then $x_1, x_2, \ldots$ have the required distribution.

Note that this procedure can be used whether or not $F(x)$ is given by a mathematical formula; for example we might take $F(x)$ to be an empirical distribution function of sample observations, possibly smoothed. When $F(x)$ is a continuous strictly increasing function the condition defining the x's may be replaced by the equation $y_1 = F(x_1)$.

In the present application we require exponentially distributed quantities with mean $\frac{3}{2}$. Here $F(x) = 1 - e^{-2x/3}$, so that we may take

$$x_1 = -\tfrac{3}{2} \log_e (1 - y_1),$$

or, alternatively,

$$x_1 = -\tfrac{3}{2} \log_e y_1,$$

since y_1 and $1 - y_1$ have the same distribution. Normally distributed quantities can be obtained similarly, but in fact special tables of these have been prepared†. The first four columns of Table 5.2 show the essential quantities entering into the calculation.

An initial condition needs to be settled at the start of each 'run' of the calculation. This may be determined from practical considerations or, if the equilibrium properties of the system are required, it may be advisable to start from what is thought to be a typical state of the process and then to discard the initial section of each run. Here we have started with the system empty.

In the present example the remainder of the calculation is straightforward, because the times at which customers start and end their service are easily calculated in sequence, and the

* M. G. Kendall and B. Babington Smith, *Tables of Random Sampling Numbers*, Cambridge, 1954.

† H. Wold, *Random Normal Deviates*, Cambridge, 1954. Also M. H. Quenouille, *Biometrika* **46** (1959), 178, has given useful short tables of random samples from other distributions.

TABLE 5.2

Outline of simulation calculation sheet for estimating mean queueing-time.

Customer number	Arrival deviation	Actual arrival time	Service-time	Admitted to server 1	Departs from server 1	Admitted to server 2	Departs from server 2	Queueing-time
1	0·4	1·4	0·2	1·4	1·6			0
2	−0·3	1·7	2·7	1·7	4·4			0
3	−0·2	2·8	1·7			2·8	4·5	0
4	0·4	4·4	4·3	4·4	8·7			0
5	0·4	5·4	1·7			5·4	7·1	0
6	−0·3	5·7	0·6			7·1	7·7	1·4
7	−0·3	6·7	2·7			7·7	10·4	1·0
8	−0·4	7·6	1·0	8·7	9·7			1·1
9	−0·5	8·5	0·6	9·7	10·3			1·2
10	0·6	10·6	3·6	10·6	14·2			0
11	−0·1	10·9	0·7			10·9	11·6	0
12	−0·1	11·9	0·9			11·9	12·8	0
. . .	. . .	. . .	. . .	. . .	. . .	. . .	. . .	. . .

queueing-times derived. (When both servers are free it is immaterial to which server the first subsequent customer is assigned.) Even in this simple example more detail would be required if, for example, it was desired to find the distribution over all time points of the number of customers in the system. It is often useful, particularly in more complicated cases, to record the evolution of the process graphically.

Table 5.2 shows the queueing-time for the first 12 customers. The calculation was continued for 200 customers and the main conclusions are given in Table 5.3.

TABLE 5.3

Properties of queueing-time estimated by simulation.

	Propn. of zero queueing-times	Mean queueing-time	Maximum queueing-time
1st 20 customers	0·90	0·02	0·2
2nd ,, ,,	0·85	0·07	0·8
3rd ,, ,,	0·60	0·35	1·8
4th ,, ,,	0·65	0·32	1·8
5th ,, ,,	0·65	0·23	1·2
6th ,, ,,	0·60	0·39	1·6
7th ,, ,,	0·30	1·18	3·1
8th ,, ,,	0·90	0·04	0·6
9th ,, ,,	0·70	0·13	1·1
10th ,, ,,	0·10	2·72	5·4
Combined	0·62	0·55	5·4

The mean queueing-time of 0·55 may be compared with results given earlier in this monograph. The analysis of section 2·4 (iii) gives the equilibrium distribution of queue-size in a system with random arrivals and otherwise the same as the present one, and from the formulae there the mean queueing-time can be shown to be 1·93. In the same way section 3.2, example (iv), gives the

distribution of queue-size at the arrival instants of the corresponding system with regular arrivals. From this the mean queueing-time
can be shown to be 0·71. It is not surprising that regular arrivals
produce much shorter queueing-times than random arrivals (with
corresponding mean inter-arrival intervals). The estimated queueing-time from the simulation study is, however, *less* than that for
the system with regular arrivals. This is certainly because of errors
of estimation; see (i) below.

The example can be used to illustrate a number of general points
connected with simulation methods.

(i) *Variability of results*

It is characteristic of many queueing systems, especially those with
very variable service-times or arrival patterns, or with high traffic
intensity, that very variable results are obtained. In the present
application long runs of zero queueing-times may occur, yet the
occurrence of two, or more, long service-times fairly close together
is enough to produce a run of appreciable queueing-times. This
means both that the estimated mean queueing-time may have a
serious random error of estimation and also that the behaviour of
the real system over a moderate time may depart appreciably from
that predicted from the equilibrium distribution. A practical consequence may well be that the system as it stands is intrinsically
unsatisfactory and needs tighter control.

To estimate the standard error of the final mean queueing-time
it would be quite wrong to use the usual statistical formula for the
standard error of the mean of independent observations; this
would ignore the strong positive correlation between the queueing-
times of adjacent customers and would seriously underestimate the
true error.

Two simple methods are available for estimating standard
errors in problems like this. The first method, which is to be pre-
ferred where practicable, is to repeat the whole procedure a number
of times independently from the beginning, calculating a mean
queueing-time from each run. The standard error of the pooled
mean can then be estimated by the usual formula. This method

requires preferably at least 10–15 runs, each sufficiently long to avoid difficulties connected with the initial conditions.

The second method which we suggest can be applied even if there is only one run. The results are divided into sections each sufficiently long for the correlation between the mean queueing-times in adjacent sections to be negligible; the choice of a suitable section length may involve trial and error. In the example the customers were, for this purpose, divided into 10 sections of 20 and a mean queueing-time found for each section. The standard error of the combined mean was estimated, with 9 degrees of freedom, to be 0·26. This standard error gives no more than a rough idea of the amount of sampling error to which the estimate is subject, because the distribution of queueing-time has a large positive skewness.

The way to decide on the length of study needed is to calculate the standard error after a preliminary investigation. If this standard error is too large, an approximate standard error can be predicted for studies of various lengths and hence a decision reached about how much further work to do.

(ii) *Method of calculation*

The calculations for the example were done using a slide-rule, and tables of random numbers and of natural logarithms. If very long runs are necessary to get adequate precision, or if results are required for many different parameter values, it will, however, be a great advantage to do the calculations on an electronic computer. It is advisable always to start with some hand calculation, because intelligent scrutiny of the detailed results may suggest an approach to a mathematical solution, or an approximation leading to a rough formula, and may also suggest what assumptions about the system are of critical importance.

When an electronic computer is used, it will be desirable to print-out at frequent intervals detailed information about the process. If the main expense of the calculation is in programing, it will usually pay to make a straightforward simulation of the process in as simple a form as possible.

K

Another method of simulation is by means of apparatus constructed specially for the particular application, e.g. by an electronic model of the system. With the increasing availability of electronic computers, it is doubtful whether the construction of special machines is often justified, although they may, of course, be useful for teaching.

(iii) *Restricted conclusions*

It is characteristic of a simulation study that each numerical result refers to one special system, whereas a mathematical formula gives us the behaviour of a range of systems. Compare the example in this section with equation (2.22), for the ratio of mean waiting-time to mean service-time in a single-server queue with random arrivals. The equation gives us not only the answer for all traffic intensities and all coefficients of variation of service-time, but also the remarkable information that moments higher than the second of the service-time distribution do not enter. To establish all this even approximately by simulation would involve a long series of trials with various traffic intensities, coefficients of variation of service-time, and shapes of service-time distribution.

When a range of conditions are to be investigated by simulation it may be helpful to use ideas from the theory of factorial experiments, including possibly fractional replication. In analysing the results of a series of such trials, it is usually a good plan not to work directly with the quantity of interest (in our example the mean queueing-time). One method is to find a theoretical solution for a system resembling the one under study as closely as possible. Each observed mean queueing-time, or whatever property is involved, is then expressed as a ratio to, or difference from, the corresponding theoretical value and these ratios or differences analysed. The point we try to make is that the true values of this ratio are more likely to vary slowly and smoothly over a wide range of parameter values than are the original true mean queueing-times; in particular, if the factorial approach is adopted, the modified values are more likely to have small interactions than are the original mean queueing-times. An alternative method is to fit an empirical formula to the mean queueing-times and then to work with the deviations from the empirical formula.

(iv) *Devices for increasing precision*

We shall not discuss in detail methods for increasing the precision of estimates derived from the type of procedure described in this section. One or two general ideas will, however, be illustrated briefly.

If the quantity of interest can be broken into components some of which can be calculated theoretically and some not, it will usually be profitable to determine by simulation only the components that cannot be found theoretically. As a rather trivial illustration of this, suppose that in the example it is required to find properties not of the queueing-time, but of the waiting-time. Instead of generating a distribution of waiting-times by adding to each queueing-time the corresponding service-time, it will be better to determine the queueing-time by simulation, and then to combine this mathematically with the exactly known distribution of service-time; in particular the estimated mean waiting time is the sum of the estimated mean queueing-time and the known mean service-time.   One fairly general method for increasing precision is by more detailed statistical analysis of the data. Suppose that the series can be divided into sections which can be treated as independent of one another, and that for each section we can find in addition to the quantity y of interest, another quantity x closely correlated with y and having a known distribution. An example, although not a very good one, would be to divide the customers in our illustrative example into sets of say 20, to let y be the mean queueing-time of the 20 customers and x their mean service-time. Then x has an exactly known distribution and it should be clear that, given a long enough series, an adjusted estimate of mean queueing-time can be found. If the relation between y and x is linear, this is roughly equivalent to correcting y by analysis of covariance. The difficulty in the present case is that the sections have to be long to be nearly independent of one another and then the correlation between y and x will be poor.

In many systems, although, strictly speaking, not in our illustrative example, the process falls naturally into sections of unequal length, behaviour in different sections being independent. For example consider any system with random arrivals, but having possibly many servers, complex systems of priority, etc. Suppose that initially there are no customers in the system. Then the subsequent behaviour can be represented like this:

$$A_1 \ldots E_1; A_2 \ldots E_2; \ldots,$$

where A_1 is the first arrival, E_1 is the next instant at which the system is empty, A_2 is the next arrival after E_1, etc. The time intervals between E_1 and A_2, E_2 and A_3, ... are independently exponentially distributed and everything that happens between A_2 and E_2 is independent of what happened between A_1 and E_1, etc.

It follows that equilibrium properties of the system can be derived from those of *tours*, a tour starting with an arrival at an empty system and ending the next time the system is again empty. By designing a sampling scheme to find appropriate properties of tours rather than of the original process, many of the Monte Carlo methods proposed for simple systems can be applied; for example weighted sampling can be used to concentrate attention on those types of tour that have greatest effect on the final answer.

5.5 Series of queues

So far we have considered systems with a single service-point and such that a customer, once he has been served, leaves the system. Quite often, however, there is a chain of service-points and a customer who has been served at one point immediately joins the input at another service-point. We shall consider here mainly a simple sequence of queues $1, \ldots, k$, say, such that a customer served at point i immediately joins the queue at point $i+1$, leaving the system only when all k stages have been completed. A modification of this arises when a customer served at point k returns immediately to point 1; we call this system a circular queue with k stages and deal with it briefly at the end of the section.

A simple sequence of k queues is often a reasonable model of an industrial process with k stages. In such applications a frequent complication is a limitation on storage space between stages. For example, if the number of customers queueing at the second stage reaches a certain level, the first stage may have to stop operation until there is room in front of the second stage for more customers.

To illustrate the theory let us consider a pair of queues ($k = 2$) and suppose moreover that they are single-server queues characterized by distribution functions of service-time $B_1(x)$, $B_2(x)$, respectively.

In extreme cases the properties of the system are easily obtained. Thus if the server of the first queue is nearly always busy, the intervals between successive departures from the first queue are independently distributed with distribution function $B_1(x)$. Hence the arrival pattern for the second queue is of the general independent type (section 1.3 (iii)); if, for example, $B_2(x)$ is exponential, the theory of section 3.2 will yield the properties of the second queue. The queueing-times, etc., in the first queue can be found from the theory for single queues and, under the conditions postulated, the queueing-times of a customer at the first queue and at the second queue will be nearly independent, so that properties of the total queueing-time can be found.

Another extreme case is when there is very rarely congestion at the first service-point. The series of departures from the first queue is then formed by displacing each arrival point by an amount equal to the corresponding service-time. If the initial arrival-pattern is random, so too will be the sequence of departure instants, and hence the properties of the second queue can be obtained from the theory for random inputs. If the initial arrivals are regular, the input into the second system will be of the 'unpunctuality' type (section 1.3 (iv)).

We now consider a problem requiring explicit consideration of both queues. Suppose that customers arrive randomly at rate α at the first queue and that the service-times in the two systems are independently exponentially distributed with means respectively $1/\sigma_1$ and $1/\sigma_2$. If $\alpha < \text{Min}(\sigma_1, \sigma_2)$, an equilibrium probability distribution exists for the numbers of customers in the two systems and can be found by the method of section 2.2.

Let $p(n_1, n_2)$ be the equilibrium probability that there are simultaneously n_1 customers in the first system and n_2 in the second, including in both cases the customer, if any, being served. Then

$$
\left.
\begin{aligned}
(\alpha + \sigma_1 + \sigma_2) p(n_1, n_2) &= \alpha p(n_1 - 1, n_2) + \sigma_1 p(n_1 + 1, n_2 - 1) \\
&\quad + \sigma_2 p(n_1, n_2 + 1) \qquad (n_1, n_2 > 0); \\
(\alpha + \sigma_2) p(0, n_2) &= \sigma_1 p(1, n_2 - 1) + \sigma_2 p(0, n_2 + 1) \\
&\qquad\qquad\qquad\qquad\qquad (n_2 > 0); \\
(\alpha + \sigma_1) p(n_1, 0) &= \alpha p(n_1 - 1, 0) + \sigma_2 p(n_1, 1) \quad (n_1 > 0); \\
\alpha p(0, 0) &= \sigma_2 p(0, 1).
\end{aligned}
\right\} \quad (22)
$$

Also the probabilities satisfy the normalizing condition

$$
\sum_{n_1, n_2} p(n_1, n_2) = 1. \tag{23}
$$

The solution of these equations may be verified to be

$$
p(n_1, n_2) = (1 - \rho_1)(1 - \rho_2) \rho_1^{n_1} \rho_2^{n_2}, \tag{24}
$$

where $\rho_i = \alpha / \sigma_i$ $(i = 1, 2)$. Therefore

(a) the distribution of the number of customers in the first system is geometric with parameter ρ_1;

(b) the distribution of the number of customers in the second system is geometric with parameter ρ_2;

(c) the distributions (a) and (b) are independent.

The result (a) follows immediately from equation (2.11) on considering the first queue by itself. The results (b) and (c) are remarkable; (b) is what we would get if the input into the second queue were random at rate α, i.e. if the departure instants from the first queue formed a Poisson process. As a matter of fact, this is indeed so, and we outline a proof.

Consider the system in statistical equilibrium and suppose that it is known that a departure from the first queue occurs at time t_0. It can be shown that the probability that the customer leaves the queue behind him empty is given by the equilibrium overall probability and is, by equation (2.11), $1 - \rho_1$. The next step towards proving the desired result is to show that if the next departure after t_0 is at $t_0 + x$, the random variable x has an exponential distribution. There are two cases to consider:

First, if the customer departing at t_0 leaves at least one other customer behind him, x has the distribution of a single service-time with Laplace transform $\sigma_1/(\sigma_1 + s)$.

Second, if the first queue is left empty, x is the sum of the interval up to the first arrival and the corresponding service-time. The Laplace transform of the distribution of this sum is

$$\sigma_1 \alpha/[(\sigma_1 + s)(\alpha + s)]$$

Therefore, the Laplace transform of the distribution of x is

$$\frac{\rho_1 \sigma_1}{\sigma_1 + s} + \frac{(1 - \rho_1)\sigma_1 \alpha}{(\sigma_1 + s)(\alpha + s)} = \frac{\alpha}{\alpha + s}, \tag{25}$$

since $\rho_1 = \alpha/\sigma_1$. The distribution of x is thus exponential with parameter α.

To complete the proof that the series of departure instants forms a Poisson process, we should show that (25) holds, given not just that a departure occurred at t_0, but given also any further information about the timing of departures before t_0. It is enough to establish that the probability that the customer departing at t_0 leaves behind an empty queue is $1 - \rho_1$, independently of occurrences before t_0, We omit a proof of this.

It can be shown that (25) holds also when the first queue is an m-server queue with random arrivals and exponential service-time.

An immediate consequence of the random input into the second queue is that to obtain properties of the waiting-time, etc., in the second queue it is not necessary to assume that the service-time there is exponentially distributed. For example, if the second queue is a single-server system with a general distribution of service-time, Pollaczek's formula (2.22) gives the mean waiting-time at the second queue. The mean time per customer spent in the whole system is the sum of the separate mean waiting-times; however, to find the variance and distribution of the total time, we need knowledge of the joint distribution of the component waiting-times. If both service-time distributions are exponential, it can be shown that the separate waiting-times are independent; hence, since, as shown in section 2.6, the separate waiting-times are exponentially distributed with parameters $\sigma_i/(1-\rho_i)$, the Laplace transform of the p.d.f. of the total time spent in the system is

$$\frac{\sigma_1}{\sigma_1+s(1-\rho_1)} \times \frac{\sigma_2}{\sigma_2+s(1-\rho_2)}$$

$$= \frac{\sigma_1\sigma_2(1-\rho_1)}{[\sigma_2(1-\rho_1)-\sigma_1(1-\rho_2)][s(1-\rho_1)+\sigma_1]}$$

$$+ \frac{\sigma_1\sigma_2(1-\rho_2)}{[\sigma_1(1-\rho_2)-\sigma_2(1-\rho_1)][s(1-\rho_2)+\sigma_2]}.$$

The corresponding probability density function is

$$\frac{\sigma_1\sigma_2}{\sigma_2(1-\rho_1)-\sigma_1(1-\rho_2)}\left[\exp\left(-\frac{\sigma_1 t}{1-\rho_1}\right)-\exp\left(-\frac{\sigma_2 t}{1-\rho_2}\right)\right].$$

These results can be extended to the equilibrium theory of a sequence of k queues with an initial random input at rate α and with service-times independently exponentially distributed with parameters $\sigma_1, \ldots, \sigma_k$. The conclusions can be summarized as follows:

(*a*) the numbers of customers in the k queues are independently distributed in the distributions of equation (2.11) corresponding to random arrivals at rate α. In particular, the distribution of the number in the ith queue is geometric with parameter $\rho_i = \alpha/\sigma_i$, if the ith queue has a single server;

(*b*) the input into the ith queue is random at rate α for all i;

(*c*) the waiting-times in the separate queues are independently distributed in the form corresponding to random arrivals at rate α. In particular, the probability density function of waiting-time in the ith queue is exponential with parameter $\sigma_i/(1-\rho_i)$, if the ith queue has a single server.

The previous results require the distributions of service-time to be exponential, except possibly in the final stage of the system. If the first queue has a single server, random input at rate α and a distribution function of service-time $B_1(x)$, we can calculate, by the argument leading to (25), the distribution of the interval between successive departures from the first queue, i.e. the distribution of the interval between successive arrivals in the second queue. For the probability that a departing customer leaves behind him an empty queue is again $1 - \rho_1$, where ρ_1 is the traffic intensity for the first queue. The rest of the argument is as before except that in (b) the Laplace transform of the sum of an arrival interval and a service-time is $\alpha B_1^*(s)/(\alpha + s)$. Therefore the Laplace transform of the distribution of departure intervals is

$$\frac{\alpha}{\alpha + s} (1 - \rho_1)B_1^*(s) + \rho_1 B_1^*(s) \tag{26}$$

If the service-time is constant with value $b_1^{(1)}$, the p.d.f. corresponding to (26) is therefore

$$(1 - \rho_1)\, \alpha\, e^{-\alpha t} + \alpha \rho_1\, e^{-\alpha(t - b_1^{(1)})}\, U(t - b_1^{(1)}),$$

where $U(x)$ is Heaviside's unit function.

Now it can be shown that different departure intervals corresponding to (26) are mutually independent only when the service-time distribution is exponential. Hence the input into the second queue cannot in general strictly be treated as of the independent interval type; we suggest, however, that a reasonable approximation to the behaviour of the second queue can usually be obtained by ignoring the dependence between different intervals and applying the theory of section 3.2.

Sometimes the first queueing system consists of several independent channels feeding into a common second stage. It is known that in such cases the statistical form of the input into the second stage is approximately random over time periods shorter than the average interval between arrivals from the same channel. This means that a random input into the second queue can often be assumed, especially when the mean queue-size is small.

We now consider briefly the effect of a limitation of storage space between stages. Even in the simple problem with two stages, each with a single server and with exponentially distributed service-times, there are several possibilities to be considered:

(a) when the number queueing at the second stage reaches a certain number, say N_2, no further customers may pass from the first to the second stage, and no customer may be served by the first server even though service of the current customer is complete. New customers arriving in the system join the first queue in the usual way;

(b) the conditions may be the same as (a) with the addition of a limit N_1 on the first queue size. If this is reached, further customers may not enter the system and are to be considered as lost;

(c) the rejection of customers waiting to enter the first queue may apply as soon as the limiting queue-size Q_2 is reached in the second queue;

(d) the limitation on storage may be expressed not in terms of the number of customers, but in terms of the total time necessary to serve all customers at the queueing point. Restrictions similar to (a) and (b) operate as soon as these cumulative service-times reach or exceed limits Q_2 and Q_1, respectively, for the two queues.

There are, no doubt, other possibilities. When the distributions of service-time are exponential, the equilibrium equations of (a), (b) and (c) can be obtained in the usual way. System (d) may sometimes be more realistic, namely when the amount of 'space' occupied by a customer is proportional to his service-time; it may also be useful sometimes as an approximation, when distributions are not exponential, as it is then sometimes easier to deal with mathematically.

To end this section we consider a circular queue, with N customers and k stages, each with a single server, the distribution of service-time at the ith stage being exponential with parameter σ_i. Let $p(n_1, \ldots, n_k)$ be the equilibrium probability that there are n_i customers at the ith stage, including the one, if any, being served.

The equilibrium probability equations are

$$\sum_{i=1}^{k} \sigma_i \epsilon(n_i) p(n_1, \ldots, n_k) = \sigma_k p(n_1 - 1, n_2, \ldots, n_k + 1)$$

$$+ \sum_{i=1}^{k-1} \sigma_i p(n_1, \ldots, n_i + 1, n_{i+1} - 1, \ldots, n_k),$$

where

$$\sum n_i = N$$

$$\epsilon(n_i) = \begin{matrix} 1 & (n_i \neq 0), \\ 0 & (n_i = 0), \end{matrix}$$

and where any $p(n_1, \ldots, n_k)$ with a negative argument is zero.

The solution to these equations is

$$p(n_1, \ldots, n_k) \propto \sigma_1^{-n_1} \sigma_2^{-n_2} \ldots \sigma_k^{-n_k}$$

$$= \lambda_1^{n_1} \lambda_2^{n_2} \ldots \lambda_k^{n_k},$$

with $\lambda_i = 1/\sigma_i$. Thus, on using the normalizing condition, we have that

$$p(n_1, \ldots, n_k) = \frac{\lambda_1^{n_1} \lambda_2^{n_2} \ldots \lambda_k^{n_k}}{\sum_{\pi} \lambda_1^{r_1} \lambda_2^{r_2} \ldots \lambda_k^{r_k}}, \tag{27}$$

where $\sum_{\pi}$ denotes summation over all sets of integers $r_1, \ldots, r_k$ such that $\sum r_i = N$.

Properties of the system can be found from (27). For example the probability that the ith stage contains no customers is

$$p_i^{(0)} = \frac{\sum_{\pi_i} \lambda_1^{r_1} \lambda_2^{r_2} \ldots \lambda_k^{r_k}}{\sum_{\pi} \lambda_1^{r_1} \lambda_2^{r_2} \ldots \lambda_k^{r_k}},$$

where $\sum_{\pi_i}$ denotes summation over all sets of integers such that $r_i = 0$ and $\sum r_i = N$. The ith server is working a proportion $1 - p_i^{(0)}$ of the time and therefore deals with $(1 - p_i^{(0)}) \lambda_i$ customers per unit time; this quantity, which is the production rate for the system, is in fact independent of i.

A special case is when the mean service-time is the same in all stages, so that the λ_i are all equal. Then

$$p_i^{(0)} = \frac{\text{number of different sets } \{r_1, \ldots, r_k\} \text{ adding up to } N, \text{ with } r_i = 0}{\text{number of different sets } \{r_1, \ldots, r_k\} \text{ adding up to } N}$$

(28)

$$= \frac{S_{k-1}^N}{S_k^N},$$

where S_k^N is the number in the denominator of (28).

Now a set of $k-1$ integers with a total less than or equal to N can always be completed into a set of k adding up to N. Therefore

$$S_k^N = S_{k-1}^N + S_{k-1}^{N-1} + \ldots + S_{k-1}^0,$$

from which we have that

$$S_k^N = \frac{(N+k-1)!}{(k-1)! N!}.$$

Thus

$$p_i^{(0)} = \frac{k-1}{N+k-1}$$

and each server is therefore working a proportion $N/(N+k-1)$ of the time. This is, in a sense, the efficiency of the system.

Koenigsberg* has given these and further results and has discussed an application to mining engineering. Clearly the situation discussed here can be generalized in numerous ways.

5.6 The busy-period distribution

To complete this chapter we return to the study of the single-server queue with random arrivals and general independent service-times, and discuss in this section the busy-periods. A few relevant elementary results were given in section 2.7.

A busy-period begins when a customer arrives to find the server free to deal with him at once (i.e. there is a 'zero' queue). A busy-period ends when the server completes the service of a customer

* E. Koenigsberg, *Operational Research Quarterly* 9 (1958), 22.

and finds that there are no customers presently demanding service (i.e. there is a 'zero' queue again). The interval between the end of one busy-period and the beginning of the next constitutes the time the server must wait for the next random arrival. The 'free-period' thus has a negative-exponential distribution with mean α^{-1}.

There are two aspects of the busy-period which can be discussed. One can consider the probability distribution of the length of the busy-period in time units. One can, alternatively, investigate the probability distribution of the number of customers served in the course of a busy-period. We shall take up both these topics, dealing first with the duration of a busy-period.

Let us suppose that the customer who 'starts' a busy-period has a service time θ. Let $\phi(t;\theta)$ be the probability density function of the duration of the busy-period started by this customer, conditional upon his service-time having the assigned value θ. Then, if we write $g(t)$ for the unconditional probability density function of the duration of a busy-period it is evident that

$$g(t) = \int\limits_0^\infty \phi(t;\theta)\,dB(\theta). \tag{29}$$

We call θ the *initial service-time* of the busy-period and say that the busy period is *generated* by θ; the function $\phi(t;\theta)$ is the p.d.f. of the busy period so generated. But the busy period generated by $\theta_1 + \theta_2$ can be considered as the sum of two random variables, corresponding to busy periods generated by θ_1 and θ_2. Thus

$$\phi(t;\theta_1+\theta_2) = \int\limits_0^t \phi(t-t';\theta_1)\,\phi(t';\theta_2)\,dt'.$$

Hence if

$$\phi^0(s;\theta) = \int\limits_0^\infty e^{-st}\phi(t;\theta)\,dt,$$

then it follows that

$$\phi^0(s;\theta_1+\theta_2) = \phi^0(s;\theta_1)\,\phi^0(s;\theta_2).$$

Thus there is a function $\kappa(s)$ such that

$$\phi^0(s;\theta) = e^{-\theta\kappa(s)}, \tag{30}$$

and our problem is to determine $\kappa(s)$. Once we have found $\kappa(s)$ we have, at once, from (29), that the Laplace transform of the busy-period distribution is

$$g^0(s) = B^\star(\kappa(s)). \tag{31}$$

The simplest way to determine $\kappa(s)$ seems to be as follows. Suppose that at the initial instant $t = 0$ a customer arrives and starts a busy-period. Let this customer have service-time θ, where θ is vanishingly small. With probability $1 - \alpha\theta$, to the first order in θ, no customer will arrive in the time interval $(0, \theta)$ and the busy-period will have a duration exactly equal to θ. With probability $\alpha\theta$, to the first order in θ, exactly one customer will arrive in the time interval $(0, \theta)$; and since θ is assumed to be arbitrarily small the busy-period will then approximate that which would have obtained if the service-time of the initial customer had been given the service-time distribution, $B(\theta)$, instead of being assumed vanishingly small. The probability of more than one customer arriving in $(0, \theta)$ is negligible compared with θ, and so such a possibility may be ignored. Thus we have that for arbitrarily small θ,

$$\phi(t, \theta) = (1 - \alpha\theta)\,\delta(t - \theta) + \alpha\theta g(t), \tag{32}$$

where $\delta(x)$ is the Dirac delta function.

If we take Laplace transforms of (32) and use (30) we find that

$$e^{-\theta\kappa(s)} = (1 - \alpha\theta)\,e^{-\theta s} + \alpha\theta g^0(s),$$

or, to the first order in θ,

$$-\theta\kappa(s) = -\alpha\theta - \theta s + \alpha\theta g^0(s).$$

Thus $$\kappa(s) = \alpha + s - \alpha g^0(s), \tag{33}$$

or, appealing to (31),

$$\kappa(s) = \alpha + s - \alpha B^\star(\kappa(s)). \tag{34}$$

Alternatively, one can also deduce from (33) and (31) that $g^0(s)$ satisfies the equation

$$g^0(s) = B^\star(\alpha + s - \alpha g^0(s)). \qquad (35)$$

It is far from obvious that (34) and (35) determine the required functions $g^0(s)$ and $\kappa(s)$ uniquely, but as a matter of fact they do, although we shall not prove this claim. It can be shown that there is only one solution of (35), for example, which for positive values of s behaves like the Laplace transform of the probability density function of a non-negative random variable.

If the traffic intensity exceeds unity then a busy-period may never end. Let us write ϖ for the probability of a busy-period ever terminating. Then evidently as s decreases through positive values to zero we expect $g^0(s)$ to increase to ϖ as a limit. Thus, from (35), we infer that ϖ must satisfy the equation

$$\varpi = B^\star(\alpha - \alpha\varpi). \qquad (36)$$

It can be shown without much difficulty that when $\rho > 1$ there is a unique solution of (36) which lies between zero and unity. This solution gives us the required probability. If $\rho \leqslant 1$ then the only possible solution is $\varpi = 1$, i.e. the busy-period will certainly terminate.

The functional equation (35) is widespread in the literature of congestion theory and is obtained in many ways. It is of somewhat limited application, however, since the derivation of $g^0(s)$ almost always involves the solution of a transcendental equation. As usual, the case where service-times have a negative-exponential distribution provides a rare tractable example. If $B^\star(s) = \sigma/(\sigma + s)$ then (35) leads to the quadratic equation

$$\alpha\{g^0(s)\}^2 - (\alpha + \sigma + s)g^0(s) + \sigma = 0. \qquad (37)$$

Letting s tend to zero leads us to the discovery that

$$\varpi = \begin{array}{ll} 1 & (\rho \leqslant 1) \\ \rho^{-1} & (\rho > 1). \end{array}$$

Thus, for this particular queueing situation, there is an especially simple relation between the traffic intensity and the probability that the busy-period will ever end. Supposing now that $\rho \leqslant 1$, we find that the appropriate solution to (37) is

$$g^0(s) = \frac{1}{2\rho}\left\{1+\rho+\frac{s}{\sigma}-\sqrt{\left[\left(1+\rho+\frac{s}{\sigma}\right)^2-4\rho\right]}\right\}.$$

The other solution of (37) can be ruled out on the grounds that we require $g^0(0)$ to equal unity. Reference to standard tables of Laplace transforms then yields the exact formula for the probability density of the busy-period distribution, namely

$$g(t) = \frac{e^{-(\sigma+\alpha)t}}{t\sqrt{\rho}}I_1(2t\sigma\sqrt{\rho}),$$

where $I_1(t)$ is the Bessel function of imaginary argument and first order.

Let us now return to the more general case for which the service-time distribution is not necessarily negative-exponential. Naturally we cannot get such precise information as we have been able to obtain for the special case. However, we can deduce the moments of $g(t)$. If we let

$$\alpha g^0(s) = \alpha+s-z, \tag{38}$$

then (35) becomes

$$s = z+\alpha[B^\star(z)-1]. \tag{39}$$

If we then write, as usual, b_r for the rth moment of the service-time distribution then (39) can be rewritten

$$s = z+\alpha\sum_{r=1}^{\infty}\frac{b_r}{r!}(-z)^r. \tag{40}$$

Thus we have expressed s as a power-series expansion in z. What we wish to deduce from (40) is the representation of z as a power-series in s. This kind of operation is referred to as the *reversion* of series and is a fairly familiar one in applied mathematics. Extensive tables are available giving the results of such a reversion. The most readily available table, which is quite adequate for our present purposes (though not the most extensive), is given in Dwight's Tables*. If we make use of these we find from (40) an expansion of the form

$$z = \sum_{r=1}^{\infty} \frac{\gamma_r}{r!}(-s)^r.$$

Substitution of this expansion in (38) then gives the expansion of $g^0(s)$ as a power-series in s. The coefficients of the various powers of s then give the moments, g_r say, of the busy-period distribution. One can by this means determine that

$$g_1 = b_1/(1-\rho),$$

$$g_2 = b_2/(1-\rho)^3,$$

$$g_3 = \frac{b_3}{(1-\rho)^4} + \frac{3\alpha b_2^2}{(1-\rho)^5},$$

$$g_4 = \frac{b_4}{(1-\rho)^5} + \frac{10\alpha b_2 b_3}{(1-\rho)^6} + \frac{15\alpha^2 b_2^3}{(1-\rho)^7}.$$

In particular, writing σ_b^2 for the variance of the service-time distribution, we have that the variance of the busy-period distribution is

$$\sigma_g^2 = \frac{\sigma_b^2 + \rho b_1^2}{(1-\rho)^3}.$$

It is gratifying, of course, to be able to obtain so much information from the seemingly unyielding equation (35). Unfortunately, it is not at all clear how valuable the moments of the busy-period distribution are in describing its general characteristics. Doubtless there will arise situations when the use of moments appears fully justified, but caution should always be exercised in their use.

We now adopt an entirely different approach to the calculation of the busy-period distribution. Suppose that the customer who starts a busy-period has a service-time θ, as before. We want to

* H. B. Dwight, *Tables of integrals and other mathematical data.* Macmillan.

L

calculate the probability that the busy-period shall terminate in $(t, t+\delta t)$ where $t > \theta$. Suppose that n customers arrive in the period $(0, t)$ and that their service-times are $S_1, S_2, \ldots, S_n$. Of course, n will be a random variable, and so will be the service-times. For the present, however, let us suppose them fixed. For the busy-period to terminate at time t we must have

$$\theta + S_1 + S_2 + \ldots + S_n = t;$$

for if the sum of the service-times is less than t then the server must become free at some time prior to t, whilst if that sum exceeds t then the busy-period cannot possibly end until some time later than t. We shall therefore suppose that this sum of service-time equals t exactly.

Even with this last supposition we are not certain that t is the first instant at which the server becomes free. If all our n customers happened to arrive after the time θ then our busy-period would end at $\theta < t$. There are other ways, too, in which the customers could arrive and produce an end to the busy-period earlier than t, even though their service-times have the required total. Our first problem is to calculate the probability that the customers, arriving at random, would in fact arrive in such a way that t is the first moment when the server becomes free. Evidently this probability depends upon $\theta, t, n, S_1, S_2, \ldots, S_n$. It may be thought that it should also depend upon α, the random-arrival rate; but this is not so, because we are assuming that exactly n customers arrive in $(0, T)$. The instants of arrival of these n customers, if they were pin-pricked on the interval $(0, t)$, have the same probability distribution as if they were n points chosen independently from a rectangular distribution on the range $(0, t)$. This is a familiar property of the random-arrivals scheme, and shows moreover that the probability we seek cannot possibly depend upon α.

Let us call this probability $P_n(\theta, t, S_1, S_2, \ldots, S_n)$. Then we shall prove the surprisingly simple result that

$$P_n(\theta, t, S_1, S_2, \ldots, S_n) = \frac{\theta}{t}. \tag{41}$$

It is to be noticed that this probability is to be shown to be independent of n and of the actual values of the service times, provided only that they sum correctly.

We shall prove (41) by the process of induction. First we ask: is it true for $n = 1$? If $n = 1$ then the only way in which the server can fail to be free at the time θ is if the one customer involved arrives before θ. The probability of this happening (if we use the rectangular distribution idea put forward above) is easily seen to be θ/t. Thus (41) is indeed true for $n = 1$.

We shall now suppose (41) has been proved for $n = 1, 2, \ldots, m-1$ and attempt to show that it is then necessarily true for $n = m$. Once we have accomplished this the truth of (41) for all n is established.

Suppose then that $n = m$. Some of these m customers will arrive before time θ, some will arrive after. Let us denote by Σ the sum of the service-times of all those customers who arrive in $(0, \theta)$, i.e. prior to θ. We shall be interested in $E\Sigma$, so we may as well calculate it before we proceed. Consider a typical one of the m customers, whose service-time happens to be S_r, say. Then S_r will contribute to the sum Σ if this customer arrives before θ, an event whose probability of occurrence is θ/T, and whose occurrence is independent of the occurrence of the corresponding events for the other $m-1$ customers. Thus it appears that

$$
\begin{aligned}
E\Sigma &= \frac{\theta}{t} S_1 + \frac{\theta}{t} S_2 + \ldots + \frac{\theta}{t} S_m \\
&= \frac{\theta}{t}(t - \theta),
\end{aligned}
\tag{42}
$$

since $\theta + S_1 + S_2 + \ldots + S_m = t$.

The server cannot be free prior to θ. Suppose we are given the value of Σ and want the conditional probability that the server cannot be free prior to T. Call this $P_m(\Sigma)$, for short. Evidently, if Σ happens to be zero then the server is free at time θ and we have

$$
P_m(0) = 0.
\tag{43}
$$

What if $\Sigma > 0$? At least one of our m customers must have arrived in $(0, \theta)$, and so there are fewer than m arrivals to occur in the interval (θ, T). For argument's sake suppose there are $k < m$ customers to arrive in this interval (θ, T), and that their service-times are $S_1', S_2', \ldots, S_k'$. Then we must have

$$\Sigma + S_1' + S_2' + \ldots + S_k' = t - \theta$$

and it appears that

$$P_m(\Sigma) = P_k(\Sigma, t - \theta, S_1', \ldots, S_k').$$

But since $k < m$ we can appeal to our assumption that (41) is proved for $n = 1, 2, \ldots, m-1$, and write

$$P_m(\Sigma) = \frac{\Sigma}{t - \theta}, \tag{44}$$

independent of k, $S_1', S_2', \ldots, S_k'$. Combining (44) and (43) we see that whether or not Σ vanishes we have

$$P_m(\Sigma) = \frac{\Sigma}{t - \theta}.$$

This is practically what we want. It is the probability we require, but conditional upon knowing Σ. To obtain the required probability we must take expectations, allowing Σ its appropriate probability distribution:

$$P_m(\theta, t, S_1, S_2, \ldots, S_n) = EP_m(\Sigma),$$

$$= \frac{E\Sigma}{t - \theta}.$$

If we refer back to (42) we see that (41) is finally proved for $n = m$. This completes the proof of (41).

The derivation of the busy-period distribution is now fairly easy. Let us write $f_n(x)$ for the probability density function of the sum of n independent service-times. Then the probability that exactly n customers arrive in $(0, t)$ and that their service-times sum to within a differential of $t - \theta$ is

$$\frac{e^{-\alpha t}(\alpha t)^n}{n!} f_n(t - \theta)\, dt.$$

Thus the probability that exactly n customers arrive in $(0, t)$ and the busy-period ends within a differential of t, is by (41), evidently

$$\frac{\theta}{t} \frac{e^{-\alpha t}(\alpha t)^n}{n!} f_n(t - \theta)\, dt.$$

We have only to sum this result over all values of n to obtain the final conclusion

$$\phi(t; \theta) = \frac{\theta}{t} \sum_{n=0}^{\infty} \frac{e^{-\alpha t}(\alpha t)^n}{n!} f_n(t - \theta). \qquad (45)$$

Notice that we have employed the convention that

$$f_0(t - \theta) = \delta(t - \theta),$$

where $\delta(x)$ is the Dirac delta function. This first term of the series refers to the possibility that no customers arrive in $(0, \theta)$, when the busy-period must necessarily end at θ.

This second treatment of the busy-period may, at first reading, appear more involved than our first treatment. However, it has the merit that it displays $\phi(t; \theta)$ explicitly as a function of θ and t and does not involve transcendental functional equations and Laplace transforms. Both approaches have their advantages, however. The first treatment is the most convenient for the derivation of moments. The result (45) is undoubtedly more convenient for the derivation of certain useful asymptotic results, as we shall now see.

The Laplace transform of $f_n(x)$ is $[f^0(s)]^n$ and for a large class of service-time distributions we have, by the Laplace Inversion Integral,

$$f_n(x) = \frac{1}{2\pi i} \int e^{sx} [f^0(s)]^n\, ds,$$

where the contour integral is along the usual path to the right of all singularities of the integrand. If we use this expression in (45) and assume that reversal of summation and integration signs can be justified then we find that

$$\phi(t;\theta) = \frac{\theta e^{-\alpha t}}{2\pi t i} \int \exp\left[-s\theta + ts + t\alpha f^0(s)\right] ds.$$

Such an integral, for large values of t, is ideally suited to approximation by the method of 'saddle points' (or 'steepest descents'). We quote here only the dominant term that can be thus obtained; and we shall assume that there is a real number ζ, to the right of all singularities of $f^0(s)$, at which $f^{0\prime}(\zeta) = -\alpha^{-1}$. Almost any problem which arises in practice satisfies this assumption. It will then necessarily follow that $f^{0\prime\prime}(\zeta) > 0$, as can be seen by twice differentiating the integral defining $f^0(s)$. The dominant term then proves to be

$$\phi(t;\theta) \sim \frac{\theta e^{-\zeta\theta}\exp\left\{-[\alpha - \zeta - \alpha f^0(\zeta)]t\right\}}{t^{3/2}[2\pi\alpha f^{0\prime\prime}(\zeta)]^{1/2}}. \tag{46}$$

The relative error involved in approximating to $\phi(t;\theta)$ in this way is $O(t^{-1})$. From this asymptotic approximation we can deduce that the probability that the busy-period terminates after time t is asymptotically

$$\frac{\theta e^{-\zeta\theta}\exp\left\{-[\alpha - \zeta - \alpha f^0(\zeta)]t\right\}}{[\alpha - \zeta - \alpha f^0(\zeta)][2\pi\alpha t f^{0\prime\prime}(\zeta)]^{1/2}}. \tag{47}$$

This formula also commits a relative error which is $O(t^{-1})$.

We relegate examples of the use of these formulae to the exercises. Nevertheless, we should point out that they are capable of wider application than immediately appears. They refer to the busy period generated by an initial service-time θ. In other words, if x_t denotes the total time needed to deal with all customers at the service-point at t, then T is the least value of t for which $x = 0$, given that $x_0 = \theta$. Reflection will show that we are posed the same problem if for any number $z > 0$ we ask ourselves: given that $x_0 = \theta + z$ what is the distribution of the time T at which we first have $x_T < z$? Thus we can use the asymptotic formulae to discuss the following situation which might occur in practice. Suppose there has been a sudden and unusual demand for service, and the work in hand has risen to some very high level θ, say. How long will elapse before the work in hand has fallen below some prescribed 'reasonable' level?

We have given first place to $\phi(t; \theta)$ rather than to $g(t)$ because we feel $\phi(t; \theta)$ to be of more practical use. However, by allowing θ to have the appropriate service-time distribution we can easily derive $g(t)$:

$$g(t) = E\phi(t; \theta).$$

Reference to (45) will show that the calculation of this expectation requires the evaluation of integrals such as

$$H_n(t) = \int_0^t \theta f_n(t - \theta) f(\theta) \, d\theta.$$

There is a neat method of doing the evaluation without recourse to transforms; nevertheless these seem to provide the simplest evaluation and so we shall adopt their use. We observe that $H_n(t)$ is the convolution of the two functions $f_n(t)$ and $tf(t)$. Thus its Laplace transform is

$$[f^0(s)]^n \left\{ -\frac{d}{ds} f^0(s) \right\} = \frac{1}{n+1} \left\{ -\frac{d}{ds} [f^0(s)]^{n+1} \right\}$$

and hence we find that $H_n(t) = tf_{n+1}(t)/(n+1)$. Using this fact we deduce from (45) that

$$g(t) = \sum_{n=0}^{\infty} \frac{e^{-\alpha t}(\alpha t)^n}{(n+1)!} f_{n+1}(t). \tag{48}$$

One can also make use of contour integrals, as before, to deduce the asymptotic approximations,

$$g(t) \sim \frac{\exp\{-[\alpha - \zeta - \alpha f^0(\zeta)]t\}}{(\alpha t)^{3/2}[2\pi f^{0''}(\zeta)]^{1/2}}, \tag{49}$$

$$\int_t^\infty g(t)\,dt \sim \frac{\exp\{-[\alpha - \zeta - \alpha f^0(\zeta)]t\}}{\alpha^{3/2}[\alpha - \zeta - \alpha f^0(\zeta)][2\pi t f^{0''}(\zeta)]^{1/2}}. \tag{50}$$

(Notice that both these results can also be inferred by giving θ the service-time distribution and taking expectations of asymptotic formulae (46) and (47)!)

Before leaving the question of the busy-period distribution in time, as opposed to its distribution in numbers of customers, there is one remark that we should make to avoid possible bewilderment. If the traffic intensity exceeds unity then we have seen that there is a distinct probability that the busy period may never end. No mention has been made of traffic intensity in our second treatment of the busy-period distribution. What interpretation, then, is to be placed on the tail-integrals like (50)? We have been careful to refer to them as the probability that the busy-period ends after t, i.e. *if it ends at all*. When $\rho > 1$ there are three possibilities: (i) busy-period ends before t; (ii) busy-period ends after t; (iii) busy-period never ends. The tail-integrals provide the probability of event (ii).

Let us now turn to a brief discussion of the distribution of the number of people served during a busy-period. The arguments are so closely parallel to those already propounded in this chapter that we shall content ourselves with little more than a sketch.

We shall write ψ_{mn} for the probability that, if we were to 'start' a busy-period with the simultaneous arrival of m customers, then exactly n customers will have been served by the time the busy-period ends, this number n to include the initial m customers who 'started' the busy-period. This probability parallels the density function $\phi(t; \theta)$ which we have already encountered, and we write

$$\Psi_m(\zeta) = \sum_{n=m}^{\infty} \psi_{mn} \zeta^n$$

for the appropriate generating function.

A 'normal' busy-period will, of course, start with the arrival of just one customer, and we write $h_n = \psi_{1n}$ for the probability that exactly n customers are served in the course of such a busy-period. Thus the probability h_n parallels the density function $g(t)$, and we write $H(\zeta)$ for the corresponding generating function.

Using these notations, we see that for $m > 1$, $n \geqslant m$,

$$\psi_{mn} = \sum_{r=m-1}^{n-1} \psi_{m-1, r} h_{n-r}$$

from which it follows easily that

$$\Psi_m(\zeta) = [H(\zeta)]^m. \tag{51}$$

This result parallels (30).

It will be recalled that in section 2.6 we encountered the distribution η_r governing the numbers of random arrivals in one service-time, and discovered an expression for its generating function in terms of the Laplace transform of the service-time distribution. It is not hard to see that for $n \geqslant 2$

$$h_n = \sum_{r=1}^{n-1} \eta_r \psi_{r, (n-1)},$$

and from this equation one can deduce that

$$H(\zeta) = \zeta \eta_0 + \zeta \eta_1 \Psi_1(\zeta) + \zeta \eta_2 \Psi_2(\zeta) + \ldots,$$

and hence that

$$\begin{aligned} H(\zeta) &= \zeta \Xi[H(\zeta)] \\ &= \zeta B^\star \{\alpha[1 - H(\zeta)]\}. \end{aligned} \tag{52}$$

This functional equation for $H(\zeta)$ parallels (35) in the treatment of the time-duration of busy-periods. By a similar reversion of series to the one we employed there, we can determine moments of the $\{h_r\}$ distribution. If we write β_r for the rth *cumulant* of the service-time distribution $B(x)$ then the first four moments of the number of customers served in the course of a busy-period turn out to be

$$\frac{1}{(1-\rho)};$$

$$\frac{1+\alpha^2\beta_2}{(1-\rho)^3};$$

$$\frac{3(1+\alpha^2\beta_2)^2}{(1-\rho)^5} - \frac{(2-\alpha^3\beta_3)}{(1-\rho)^4};$$

$$\frac{(6+\alpha^4\beta_4)}{(1-\rho)^5} - \frac{10(1+\alpha^2\beta_2)(2-\alpha^3\beta_3)}{(1-\rho)^6} + \frac{15(1+\alpha^2\beta_2)^3}{(1-\rho)^7}.$$

The variance of the number of customers served during a busy-period is thus, writing σ_b^2 in place of β_2,

$$\frac{\rho+\alpha^2\sigma_b^2}{(1-\rho)^3}$$

There is also a treatment of the distribution $\{h_r\}$ on the lines of our second treatment of $g(t)$. In the present case the results are extremely simple. Let us write $\eta_n^{(r)}$ for the probability that exactly n random arrivals occur during r independent service-times. Thus if the probabilities $\{\eta_n^{(r)}\}$ have a generating function $\varXi_{(r)}(\zeta)$ then

$$\varXi_{(r)}(\zeta) = [\varXi(\zeta)]^r$$
$$= \{B^\star[\alpha(1-\zeta)]\}^r.$$

One can then prove that

$$\psi_{mn} = \frac{m}{n}\eta_{n-m}^{(n)}, \tag{53}$$

and thence by putting $m = 1$ that

$$h_n = \frac{1}{n}\eta_{n-1}^{(n)}. \tag{54}$$

These results are much less complicated than the corresponding results (45) and (48).

To obtain asymptotic formulae for ψ_{mn} (and h_n) for large values of n we need obtain asymptotic approximations for $\eta_{n-m}^{(n)}$. It can be shown that

$$\eta_{n-m}^{(n)} = \frac{1}{2\pi i} \int_{-i\infty}^{i\infty} \frac{[B^\star(z)]^n}{\left(1 - \dfrac{z}{\alpha}\right)^{n-m+1}} \frac{dz}{\alpha},$$

and this contour integral also is amenable to attack by the method of 'saddle-points'.

For practically every service-time distribution one is liable to be interested in, there is a negative number $-\epsilon$, say, such that as z decreases through real values to $-\epsilon$, the Laplace transform $B^\star(z)$ increases without bound (sometimes $\epsilon = \infty$). The point $z = -\epsilon$ is a singularity of $B^\star(z)$. For such service-time distributions it is easy to show that the function

$$\frac{B^\star(z)}{\left(1 - \dfrac{z}{\alpha}\right)} = e^{\psi(z)}, \text{ say}$$

has a unique minimum λ in the interval $(-\epsilon, \alpha)$. If we apply the method of 'saddle-points' we find that, for large values of n,

$$\eta_{n-m}^{(n)} \sim \frac{1}{\alpha[2\pi n\psi''(\lambda)]^{1/2}} \frac{[B^\star(\lambda)]^n}{\left(1 - \dfrac{\lambda}{\alpha}\right)^{n-m+1}}. \tag{55}$$

This approximation, used in formulae (53) and (54), provide valuable approximations to ψ_{mn}, h_n.

Finally, let us notice that ψ_{mn} is not only the probability that a server faced with a queue of size m will have to serve n customers before he is next free. It is also the probability that if the queue has length $m + M$, say, then n customers will be served before the queue sinks to size M. Thus we can use ψ_{mn} in discussing how long must elapse before a very long queue will disperse.

APPENDIX I

Bibliographical Notes

General

The main purpose of the following notes is to indicate key papers on which the treatment in this monograph is based and from which references to other work may be obtained. Doig (1957) has given a bibliography of publications on queueing and related fields up to 1957. Current work is abstracted in *Mathematical Reviews* and in *International Journal of Abstracts: Statistical Theory and Method*. New British and American work appears mostly in the following journals: *Ann. Math. Statist.*, *Bell System Tech. J.*, *Biometrika*, *J. R. Statist. Soc.* B, *Operational Research Quarterly*, and *Operations Research*.

Chapter I: Introduction

The earliest systematic mathematical work on queueing problems is that of Erlang, of the Copenhagen Telephone Company, whose first paper on congestion in telephone exchanges was written in 1909. Jensen (1948) has reviewed Erlang's work from a modern point of view. The text-book on probability theory by Fry (1928) discusses a number of telephone problems. Other notable early work is that of Pollaczek, Khinchine, and Palm (1943). Pollaczek's and Khinchine's work is most accessible through their books (Pollaczek, 1957; Khinchine, 1960).

Much recent mathematical work stems from two important papers by D. G. Kendall (1951, 1953). Kendall introduced also the classification of queues by input, service mechanism and queue-discipline.

Chapter II: Some simple queues with random arrivals

Much effort has gone into proving the existence of limiting probability distributions for queueing processes. The reader who wishes to pursue this subject should read first the account of Markov Chains given by Feller (1957). He should then read the papers of D. G. Kendall (1951, 1953), Lindley (1952), and Kiefer and Wolfowitz (1955). Smith (1955) has given useful general theorems for discussing a large class of queueing situations.

The account of the single-server queue with random arrivals in section 2.6 is based on the work of D. G. Kendall (1951).

In his book, Morse (1958) applied the methods of sections 2.2–2.4 to a wide variety of special queueing problems.

Chapter III: More about simple queues

The derivation of the transient probability distribution of queue-size, given in section 3.1, is new. It may be compared with the paper of Bailey (1954). Very much more general results concerning the transient behaviour of many-server queues with random arrivals and exponential service-times have been given by Karlin and McGregor (1958). The asymptotic discussion of the rush-hour problem is based on a treatment by Cox (1955b).

On queues with many servers the main reference is D. G. Kendall (1953); see also Winsten (1959).

The first results on queues with non-preemptive priorities appear to be due to Cobham (1954). A very detailed and rigorous account has been given by Kesten and Runnenburg (1957). The treatment of section 3.3 is new.

Preemptive priority is discussed by White and Christie (1958) and by Stephan (1958).

Chapter IV: Machine interference

The first mathematical discussion of machine interference seems to be due to Khinchine (1933), some other early papers being those of Kronig and Mondria (1943) and Palm (1943). Ashcroft (1950) gave a useful table for a system with constant service-time. Benson and Cox (1951) discussed, among other things, the problem of ancillary work, the effect on efficiency of changes in the distribution of service-time, and the effect of priorities. Benson (1957) has considerably extended this work. Peck and Hazelwood (1958) have given a very extensive table for multi-server problems.

Mack, Murphy, and Webb (1957), Howie and Shenton (1959), and Ben-Israel and Naor (1960) have considered regular systems of patrolling.

A number of mathematically incorrect solutions of the machine interference problem have been published.

Chapter V: More specialized topics

The method of stages was introduced first by Erlang; see Jensen (1948). D. G. Kendall (1948) employed the method in a notable paper on the growth of bacteria, and it has since been used by many writers on queueing and allied problems. An extension of the technique, using complex-valued transition probabilities, has been given by Cox (1955a).

The formulation of the integral equation of section 5.3 is due to Lindley (1952). A detailed study of the solution was made by Smith (1953). Kiefer and Wolfowitz (1955) obtained the integral equation of the many-server queue. Attention is drawn also to the paper by Spitzer (1957) on Wiener–Hopf equations.

Simulation and Monte Carlo methods have been extensively applied to queueing problems but little seems to have been published about the general aspects of applications such as those of section 5.4.

Some useful papers on series of queues are those of Jackson (1956), Reich (1957), and Finch (1959); references to other work can be obtained from Reich's and Finch's papers.

The distribution of the number of customers served in a busy-period was discussed by Borel (1942); the functional equation (35) for the Laplace–Stieltjes transform of the distribution of the duration of a busy period was first obtained by D. G. Kendall (1951). The arguments used in section 5.6 are, however, new.

References

ASHCROFT, H. (1950). 'The productivity of several machines under the care of one operator', *J. R. Statist. Soc.* B, **12**, 145–151.

BAILEY, N. T. J. (1954) 'A continuous time treatment of a simple queue using generating functions', *J. R. Statist. Soc.* B, **16**, 288–291.

BEN-ISRAEL, A., and NAOR, P. (1960). 'A problem of delayed service, I, II', *J. R. Statist. Soc.* B, **22**, 245–276.

BENSON, F. (1957). *Ph.D. Thesis*, University of Birmingham.

BENSON, F., and COX, D. R. (1951). 'The productivity of machines requiring attention at random intervals', *J. R. Statist. Soc.* B, **13**, 65–82.

BOREL, E. (1942). 'Sur l'emploi du théorème de Bernoulli pour faciliter le calcul d'une infinité de coefficients. Application au problème de l'attente à un guichet', *Comptes Rendus Acad. Sci. Paris*, **214**, 452–456.

COBHAM, A. (1954). 'Priority assignment in waiting line problems', *J. Opns. Res. Soc. Amer.*, **2**, 70–76. Also: 'Priority assignment—a correction', *J. Opns. Res. Soc. Amer.* (1955), **3**, 547.

COX, D. R. (1955a). 'A use of complex probabilities in the theory of stochastic processes', *Proc. Camb. Phil. Soc.*, **51**, 313–319.

COX, D. R. (1955b). 'The statistical analysis of congestion', *J. R. Statist. Soc.* A, **118**, 324–335.

DOIG, A. (1957). 'A bibliography on the theory of queues', *Biometrika*, **44**, 490–514.

FELLER, W. (1957). *An Introduction to Probability Theory and its Applications*. Second edition. New York: Wiley.

FINCH, P. (1959). 'The output process of the queueing system M/G/1', *J. R. Statist. Soc.* B, **21**, 375–380.

FRY, T. C. (1928). *Probability and its Engineering Uses*. New York: van Nostrand.

HOWIE, A. J., and SHENTON, L. R. (1959). 'The efficiency of automatic winding machines with constant patrolling time', *J. R. Statist. Soc.* B, **21**, 381–395.

JACKSON, R. R. P. (1956). 'Random queueing processes with phase-type service', *J. R. Statist. Soc.* B, **18**, 129–132.

JENSEN, A. (1948). 'An elucidation of A. K. Erlang's statistical works through the theory of stochastic processes', pp. 23–100 of the memoir of Brockmeyer, Halstrom, and Jensen, *The Life and Works of A. K. Erlang*. Copenhagen.

KARLIN, S., and MCGREGOR, J. (1958). 'Many server queueing processes with Poisson input and exponential service times', *Pacific J. Math.*, **8**, 87–118.

KENDALL, D. G. (1948). 'On the role of variable generation time in the development of a stochastic birth process', *Biometrika*, **35**, 316–330.

KENDALL, D. G. (1951). 'Some problems in the theory of queues', *J. R. Statist. Soc.* B, **13**, 151–185.

KENDALL, D. G. (1953). 'Stochastic processes occurring in the theory of queues and their analysis by means of the imbedded Markov Chain', *Ann. Math. Statist.* **24**, 338–354.

KESTEN, H., and RUNŃENBERG, J. TH. (1957). 'Priority in waiting line problems', *Proc. Akad. Wet. Amst.* A, **60**, I, 312–324; II, 325–336. (*Indagationes Math.*, **19**, 1957).

KHINCHINE, A. JA. (1933). 'Uber die mittlere Dauer des Stillstandes von Maschinen', *Mat. Sbornik*, **40**, 119–123 (Russian; German summary).

KHINCHINE, A. JA. (1960). *Mathematical Methods in the Theory of Queueing* (translated by Andrews, D. M., and Quenouille, M. H.). London: Griffin.

KIEFER, J., and WOLFOWITZ, J. (1955). 'On the theory of queues with many servers', *Trans. Amer. Math. Soc.*, **78**, 1–18.

KRONIG, R. (1943). 'On time losses in machinery undergoing interruptions', Part I, Part II (with Mondria, H.), *Physica*, **10**, 215–224, 331–336.

LEDERMANN, W., and REUTER, G. E. H. (1954). 'Spectral theory for the differential equations of simple birth and death processes', *Phil. Trans. Roy. Soc.* A, **246**, 321–369.

LINDLEY, D. V. (1952). 'The theory of queues with a single server', *Proc. Camb. Phil. Soc.*, **48**, 277–289.

MACK, C., MURPHY, T., and WEBB, N. L. (1957). 'The efficiency of N machines uni-directionally patrolled by one operator when walking times and repair times are constants', *J. R. Statist. Soc.* B, **19**, 166–172.

MORSE, P. M. (1958). *Queues, Inventories, and Maintenance.* New York: Wiley.

PALM, C. (1943). 'Intensitatschwankungen im Fernsprechverkehr', *Ericsson Technics*, **6**, 1–189.

PECK, L. G., and HAZELWOOD, R. N. (1958). *Finite Queueing Tables.* New York: Wiley.

POLLACZEK, F. (1957). *Problemes stochastiques posés par le phenomène de formation d'une queue d'attente à un guichet et par des phénomènes apparentés. Mémor. Sci. Math. no.* 136. Paris: Gauthier-Villars.

REICH, E. (1957). 'Waiting times when queues are in tandem', *Ann. Math. Statist.* **28**, 768–773.

SMITH, W. L. (1953). 'On the distribution of queueing times', *Proc. Camb. Phil. Soc.*, **49**, 449–461.

SMITH, W. L. (1955). 'Regenerative stochastic processes'. *Proc. Roy. Soc.* A, **232**, 6–31.

SPITZER, F. (1957). 'The Wiener–Hopf equation whose kernel is a probability density', *Duke Math. J.*, **24**, 327–343.

STEPHAN, F. F. (1958). 'Two queues under preemptive priority with Poisson arrival and service rates', *Operations Research*, **6**, 399–418.

WINSTEN, C. B. (1959). 'Geometric distributions in the theory of queues', *J. R. Statist. Soc.* B, **21**, 1–35.

WHITE, H., and CHRISTIE, L. S. (1958). 'Queueing with preemptive priorities or with breakdown', *Operations Research*, **6**, 79–95.

M

APPENDIX II

Exercises and Further Results

=====

1. Suppose that in the scheme of section 2.3 each 'arrival' is actually the arrival of two customers. Show that this leads to a modification of the equilibrium equations in which

$$(\alpha + \sigma)p_n = \alpha p_{n-2} + \sigma p_{n+1} \quad (n \geqslant 2)$$

and write down also the equations for $n = 0, 1$. Prove that if $\rho = 2\alpha/\sigma$ the generating function of $\{p_n\}$ is

$$P(\zeta) = \frac{2(1-\rho)(1-\zeta)}{\rho\zeta^3 - (2+\rho)\zeta + 2}$$

and use this to obtain the mean and variance of queue-size.

[section 2.3]

2. Repeat Exercise 1 with the figure two replaced by an arbitrary integer m. Show that now the generating function is, with $\rho = m\alpha/\sigma$,

$$P(\zeta) = \frac{m(1-\rho)(1-\zeta)}{\rho\zeta^{m+1} - (m+\rho)\zeta + m}.$$

[section 2.3]

3. Show that the results of Exercise 2 can be reinterpreted as follows. Customers arrive singly at random, and the service-time distribution is of the special Erlangian type with parameter m, so that service can be regarded as the sum of m independent exponentially distributed quantities. The state variable is the total number of stages of service still to be completed. Show that the formula of Exercise 2 leads to a mean queue-size in agreement with the general formula (20) of section 2.6.

[sections 2.3, 5.2]

4. Prove by the argument of section 2.7 that in the system of section 2.4 with arrival and service rates dependent on queue-size, the mean busy period is $(S-1)/\alpha_0$.

[sections 2.4, 2.7, 5.6]

5. Travellers and taxis arrive at a service-point independently at random at rates α, β. Let the queue-size at time t be q_t, a negative value denoting a line of taxis, a positive value a queue of travellers. Show that, starting with $q_0 = 0$, the distribution of q_t is given by the difference between independent Poisson variables of means αt, βt. Show by using the normal approximation to the Poisson distribution that if $\alpha = \beta$, the probability that $-k \leqslant q_t \leqslant k$ is, for large t, $(2k+1)(4\pi\alpha t)^{-1/2}$. List some of the effects which, in practice, would probably make the mathematical model inadequate.

[section 2.4]

6. Suppose that m operators attend N machines, that stops occur randomly in running-time at rate α, that any stop is attended as soon as possible by any free operator and that service-time is exponentially distributed with mean $1/\sigma$. Obtain the equilibrium equations and prove that if p_n is the probability that n machines are stopped

$$p_n = \frac{N!}{(N-m)!\,(m-1)!\,m^{n-m+1}}\gamma^n p_0 \quad (n \geqslant m)$$

$$= \frac{N!}{(N-n)!\,n!}\gamma^n p_0 \quad (n < m),$$

where $\gamma = \alpha/\sigma$.

[sections 2.4, 4.3]

7. Suppose that each machine is liable to m types of stop and that the m servers each specialize on one type of stop, the ith stop occurring randomly at rate α_i and having service-time exponentially distributed with mean $1/\sigma_i$. Show that the equilibrium probability that there are n_i machines stopped from type i $(i = 1,\ldots,m)$ is

$$p(n_1,\ldots,n_m) = \frac{N!}{(N-\sum n_i)!}\, \Pi\gamma_i^{n_i} p_0,$$

where p_0 is the probability that all machines are running, and $\gamma_i = \alpha_i/\sigma_i$.   [sections 2.4, 4.3]

8. Show that if $B(x)$ is a distribution function with mean b_1 and if

$$\hat{B}(x) = \int_0^x \frac{1-B(z)}{b_1}\, dz,$$

then $\hat{B}(x)$ is a distribution function, and

$$\hat{B}^\star(s) = \frac{1-B^\star(s)}{b_1 s}.$$

Show that in the model and notation of section 2.6, with arrivals random,

$$V^\star(s) = (1-\rho) \sum_{n=0}^{\infty} \rho^n[\hat{B}^\star(s)]^n.$$

Writing $\hat{B}_n(x)$ for the n-fold convolution of $\hat{B}(x)$ with itself, deduce that

$$V(x) = (1-\rho) \sum_{n=0}^{\infty} \rho^n \hat{B}_n(x).$$

Note that the function $\hat{B}_0(x)$ is defined to be $U(x)$, the Heaviside Unit Function, so that the first term corresponds to the probability $1-\rho$ of finding the server free.

[section 2.6]

9. Suppose all service-times equal exactly two time units. Show that the function $\hat{B}(x)$ of Exercise 8 represents a rectangular distribution. Let $\rho = 0.2$ and demonstrate how the result of Exercise 8 may be used to provide a fair approximation to the queueing-time distribution.

[section 2.6]

10. Suppose that arrivals are of the aggregated random type (section 1.3 (v)) with arrival instants occurring randomly at rate α and with a probability g_r for there being r customers arriving together at an arrival instant. Show that the probability generating function of the number of customers arriving in time t is

$$\exp\left[\alpha t G(\zeta) - \alpha t\right],$$

where $G(\zeta) = \sum g_r \zeta^r$. Show further that the probability generating function of the random variables ζ_n of section 2.6, the number of arrivals during the service of a customer, is $B^\star[\alpha t - \alpha t G(\zeta)]$.

[section 2.6]

11. Show that the argument of section 2.6 can be used to obtain the mean and probability generating function of queue-size for the system of Exercise 10. Prove that the ratio of the mean queueing-time, from the arrival of a group of customers to the start of the service of the group, to the mean service-time of a customer, is

$$\frac{\rho \mu_r}{2(1-\rho)}\left\{1 + \frac{C_b^2}{\mu_r} + C_r^2\right\},$$

where μ_r is the mean group size, C_r the coefficient of variation of group size, and $\rho = \mu_r b_1 \alpha$. If the customers within a group arriving together are served in random order, show that the mean queueing-time for a single customer to mean service-time is the above increased by $\frac{1}{2}\mu_r(1 + C_r^2) - \frac{1}{2}$.

[section 2.6]

12. Customers arrive randomly at rate α. Service is instantaneous but is only available at service instants, the intervals between successive service instants being independently distributed with distribution function $C(x)$; the maximum number of customers that can be served at any service instant is m. Show that if q_n is the number of customers in the system just before the nth service instant, then

$$q_{n+1} = \begin{cases} q_n + x_n - m & (q_n \geq m) \\ x_n & (q_n < m), \end{cases}$$

where x_n is the number of arrivals in the interval between the nth and $(n+1)^{\text{th}}$ service instants. Prove that the probability generating function of x_n is $C^\star(\alpha - \alpha\zeta)$. Hence show that the equilibrium probability generating function of q_n, $\Pi(\zeta)$, say, is

$$\Pi(\zeta) = \frac{\displaystyle\sum_{j=0}^{m-1} \pi_j(\zeta^m - \zeta)}{\zeta^m\{C^\star(\alpha - \alpha\zeta)\}^{-1} - 1},$$

where $\pi_i = \text{prob}(q_n = i) \quad (i = 0, \ldots, m-1)$.

The π_i can be determined from the condition that within the unit ζ-circle, the numerator must vanish when the denominator does. Hence show that if $C(x) = e^{-\mu x}$,

$$\Pi(\zeta) = (\zeta_m - 1)/(\zeta_m - \zeta),$$

where ζ_m is the zero of $\zeta^m\{1 + \alpha(1 - \zeta)/\mu\} - 1$ outside the unit circle.*

[section 2.6]

13. Suppose that customers arrive randomly, that there is a single server, and that the distribution of service-time is exponential. Suppose also that customers are divided into successive 'parties' of k, and that a customer can enter the service-point only when all members of his party have arrived. Show that the properties of this system can be obtained by the device of section 5.2.

[section 5.2]

* See N. T. J. Bailey, *J. R. Statist. Soc.* B, **16** (1954), 80, and F. Downton, *J. R. Statist. Soc.* B, **17** (1955), 256 and **18** (1956), 265, for detailed analysis of this system.

14. Show, in the notation of section 5.3, that if $b^0(s)$ is the reciprocal of a polynomial of degree l, so too is the Laplace transform of the distribution of waiting time.

[section 5.3]

15. For the remaining exercises the notation of section 5.6 is used. Suppose that service-time has a gamma distribution, with

$$B^\star(s) = \frac{\sigma^m}{(\sigma+s)^m}.$$

Show that

$$\zeta = -\sigma(1-\rho^{1/(m+1)});$$

and then deduce that the tail of the busy-period distribution is asymptotically

$$g(t) \sim \frac{1}{\alpha t^{3/2}}\left\{\frac{\sigma\rho^{1/(m+1)}}{2\pi(m+1)}\right\}^{1/2} e^{-Kt\sigma},$$

where $K = 1 - (m+1/m)\rho^{1/(1+m)} + \rho/m$.

[section 5.6]

16. Suppose that a rush-hour has suddenly increased the waiting-time to a level $\theta + \theta_0$. Show that the probability that a period longer than t will elapse, before waiting-times reduce to θ_0 or less, is given asymptotically by

$$\frac{\theta}{K}\left\{\frac{\rho^{1/(1+m)}}{2\pi(1+m)\,\sigma t}\right\}^{1/2} \exp\left\{\theta(1-\rho^{1/(1+m)}) - \sigma Kt\right\}.$$

[section 5.6]

17. By putting $m = 1$ in the results of exercises 15 and 16, deduce the results appropriate when service-times have a negative exponential distribution.

[section 5.6]

18. Suppose that service-time is equal to b_1 say. Show that

$$B^\star(s) = e^{-b_1 s}$$

and hence that $\zeta = b_1^{-1} \log \rho$. Deduce that for large t the tail of the busy-period distribution is given asymptotically by

$$g(t) \sim \frac{e^{-\alpha t[1-(1/\rho)]} \rho^{(\alpha t/\rho)}}{\alpha t^{3/2}(2\pi b_1)^{1/2}}.$$

Show that this can be deduced from the result of Exercise 16 by putting $\sigma = m/b_1$ and letting $m \to \infty$.

[section 5.6]

19. Show that in Exercise 18 the probability that exactly n customers are served in a busy-period is asymptotically

$$h_n \sim \frac{\rho^{n-1} e^{n(1-\rho)}}{n^{3/2}(2\pi)^{1/2}}.$$

[section 5.6]

Notes on the theory of probability and Laplace–Stieltjes transforms

Chapter I of this monograph is mostly about generalities and little technical knowledge is assumed in it. The later chapters, however, do require familiarity with the formal rules of the theory of probability and with certain elementary properties of Laplace–Stieltjes transforms. We make no attempt to give a potted account of the details of these matters here; such condensations tend to be incomprehensible even to those quite familiar with the subjects. For a good grounding in probability theory we recommend study of *Introduction to Probability Theory and its Applications*, vol. 1, by W. Feller (New York: Wiley, 2nd ed., 1957), and for a detailed account of Laplace–Stieltjes transforms we recommend the early chapters of *The Laplace Transform*, by D. V. Widder (Princeton University Press, 1946).

The following remarks are intended as a very brief introduction to the use of Laplace–Stieltjes transforms in probability theory. If we have a continuous random variable X, with a probability density function $f(x)$, the mathematical expectation of a function $\phi(X)$ is defined in the usual way to be

$$E\phi(X) = \int_{-\infty}^{\infty} \phi(x) f(x) \, dx. \qquad (1)$$

If we have a discrete random variable X, taking values $x_0, x_1, \ldots$ with probabilities $f_0, f_1, \ldots$, the mathematical expectation is now defined as

$$E\phi(X) = \sum_i \phi(x_i) f_i. \qquad (2)$$

(Very often $x_0 = 0$, $x_1 = 1$, $x_2 = 2$, ...)

It is convenient to subsume the formulae (1) and (2) into one, by introducing the cumulative distribution function $F(x)$ of the random variable, namely

$$F(x) = \text{prob}(X \leqslant x),$$

and by establishing the notation

$$E\phi(X) = \int\limits_{-\infty}^{\infty} \phi(x)\,dF(x); \qquad (3)$$

such an integral is called a Stieltjes integral.*

In the continuous case $F(x)$ is the integral of $f(x)$, so that $F'(x) = f(x)$; (3) then reduces to (1). In the discrete case, $F(x)$ is a step function which jumps by f_i at $x = x_i$ $(i = 0, 1, 2, \ldots)$ and is otherwise constant; (3) is then defined to be the sum (2). If we have a random variable, such as a waiting-time, whose distribution is partly continuous and partly discrete, the notation (3) can still be used in a natural way.

Most of the random variables considered in this monograph take only non-negative values, and then the lower limit of integration in (3) can be taken as zero. We do this from now on. If we take the special case $\phi(x) = e^{-sx}$, we have for Ee^{-sX} the expression

$$\int\limits_{0}^{\infty} e^{-sx}\,dF(x) = F^\star(s), \qquad (4)$$

say. We call (4) the (one-sided) *Laplace–Stieltjes transform* of $F(x)$. In the special case of a continuous distribution (4) becomes

$$\int\limits_{0}^{\infty} e^{-sx} f(x)\,dx = f^0(s), \qquad (5)$$

say. This is called the (ordinary) *Laplace transform* of $f(x)$, and is a function widely used in applied mathematics.

The following properties of $F^\star(s)$ are important:

(i) $F^\star(s)$ uniquely determines the distribution function $F(x)$;

(ii) if X_1, X_2 are independent random variables, with distribution functions $F_1(x)$, $F_2(x)$, the Laplace–Stieltjes transform of the distribution function of $X_1 + X_2$ is $F_1^\star(s) F_2^\star(s)$. For the required transform is

$$
\begin{aligned}
Ee^{-sX} &= E[e^{-sX_1}.e^{-sX_2}] \\
&= Ee^{-sX_1}.Ee^{-sX_2} \\
&= F_1^\star(s) F_2^\star(s),
\end{aligned}
$$

the second line following because the expectation of the product of independent random variables is the product of their expectations. The result extends to the sum of n independent random variables;

(iii) if $F^\star(s)$ can be expanded in a series of powers of s the coefficients determine the moments of X. In fact, except for the sign of s, (5) is the function commonly called the moment generating function.

One particular result that is used quite frequently is that the Laplace transform of the probability density function

$$
\frac{\alpha(\alpha x)^{k-1} e^{-\alpha x}}{(k-1)!} \quad (x \geqslant 0) \tag{6}
$$

is $\alpha^k/(\alpha+s)^k$.

As a simple example of the application of these ideas, the reader is advised to examine the argument in section 1.3 (ii), leading to the result in equation (11) that when arrivals are random at rate α, the time from a fixed point up to the kth following arrival has the distribution (6).

Author Index

Subject Index